BIG IDEAS MATH.
Algebra 1

Assessment Book

- Prerequisite Skills Test
- Pre-Course Test with Item Analysis
- Quiz
- Chapter Tests
- Alternative Assessment
- Performance Task
- Cumulative Test
- Post Course Test with Item Analysis

Erie, Pennsylvania

Photo Credits
Cover Image Allies Interactive/Shutterstock.com

Copyright © by Big Ideas Learning, LLC. All rights reserved.

Permission is hereby granted to teachers to reprint or photocopy in classroom quantities only the pages or sheets in this work that carry a Big Ideas Learning copyright notice, provided each copy made shows the copyright notice. These pages are designed to be reproduced by teachers for use in their classes with accompanying Big Ideas Learning material, provided each copy made shows the copyright notice. Such copies may not be sold and further distribution is expressly prohibited. Except as authorized above, prior written permission must be obtained from Big Ideas Learning, LLC to reproduce or transmit this work or portions thereof in any other form or by any other electronic or mechanical means, including but not limited to photocopying and recording, or by any information storage or retrieval system, unless expressly permitted by copyright law. Address inquiries to Permissions, Big Ideas Learning, LLC, 1762 Norcross Road, Erie, PA 16510.

Big Ideas Learning and *Big Ideas Math* are registered trademarks of Larson Texts, Inc.

Printed in the United States

ISBN 13: 978-1-60840-855-9
ISBN 10: 1-60840-855-8

56789-VLP-18 17

Contents

About the Assessment Book 1

Prerequisite Skills Test.. 2

Pre-Course Test with Item Analysis 8

Chapter 1   Solving Linear Equations... 14
Chapter 2   Solving Linear Inequalities 26
Chapter 3   Graphing Linear Functions 36
Chapter 4   Writing Linear Functions 52
Chapter 5   Solving Systems of Linear Equations 64
Chapter 6   Exponential Functions and Sequences 80
Chapter 7   Polynomial Equations and Factoring 92
Chapter 8   Graphing Quadratic Functions 104
Chapter 9   Solving Quadratic Equations 122
Chapter 10 Radical Functions and Equations 134
Chapter 11 Data Analysis and Displays 144

Post Course Test with Item Analysis 160

Answers A1

About the Assessment Book

Prerequisite Skills Test with Item Analysis

The Prerequisite Skills Test checks students' understanding of previously learned mathematical skills they will need to be successful in Algebra 1. The Item Analysis can be used to determine topics that need to be reviewed.

Pre-Course Test with Item Analysis

Post Course Test with Item Analysis

The Pre-Course Test and Post Course Test cover key concepts that students will learn in their Algebra 1 course. The Item Analysis can be used to determine topics that need to be reviewed.

Quiz

The Quiz provides ongoing assessment of student understanding. The quiz appears at the halfway point of the chapter.

Chapter Tests

The Chapter Tests provide assessment of student understanding of key concepts taught in the chapter. There are two tests for each chapter.

Alternative Assessment with Scoring Rubric

Each Alternative Assessment includes at least one multi-step problem that combines a variety of concepts from the chapter. Students are asked to explain their solutions, write about the mathematics, or compare and analyze different situations.

Performance Task

The Performance Task presents an assessment in a real-world situation. Every chapter has a task that allows students to work with multiple standards and apply their knowledge to realistic scenarios.

Cumulative Test

The Cumulative Test provides students practice answering questions in standardized test format. The assessments cover material from multiple chapters of the textbook. The questions represent problem types and reasoning patterns frequently found on standardized tests.

Name _____ Date _____

Skills Test: Prerequisite Skills Test

Add or subtract.

1. $-9 + (-15)$
2. $2 + (-3)$
3. $6 - 9$
4. $-6 - 11$
5. $13 + 8$
6. $-12 - (-10)$

Multiply or divide.

7. $2(-7)$
8. $-8 \cdot 2$
9. $9 \div 3$
10. $25 \div (-5)$
11. $-30 \div (-6)$
12. $-1(-7)$

Solve the problem and specify the units of measure.

13. The length of a rectangle is 6 feet and the width is 3 feet. Find the perimeter of the rectangle.

14. One side of a square measures 9 centimeters. Find the area of the square.

Graph the number.

15. 4

16. $|-3|$

17. $-6 + |5|$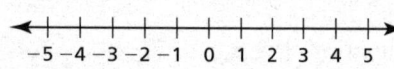

18. $1 - |-3|$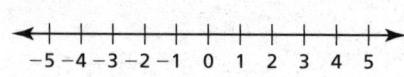

Complete the statement with <, >, or =.

19. 3 ____ 7
20. -1 ____ 4
21. -4 ____ -10
22. $|-6|$ ____ -3

Evaluate the expression for the given value of x.

23. $2x - 6$; $x = 9$
24. $-7 + 9x$; $x = 3$
25. $12x + 13$; $x = 5$

Answers

1. _____
2. _____
3. _____
4. _____
5. _____
6. _____
7. _____
8. _____
9. _____
10. _____
11. _____
12. _____
13. _____
14. _____
15. _See left._
16. _See left._
17. _See left._
18. _See left._
19. _____
20. _____
21. _____
22. _____
23. _____
24. _____
25. _____

Name_____ Date_____

Skills Test: Prerequisite Skills Test (continued)

Evaluate the expression for the given value of x.

26. $-x - 12;\ x = 4$ 27. $13 - 7x;\ x = -10$ 28. $11x + 17;\ x = -6$

Plot the point in the coordinate plane. Describe the location of the point.

29. $A(4, 2)$ 30. $B(-1, 3)$ 31. $C(-5, -3)$ 32. $D(3, 0)$

Use the graph to answer the question.

33. Which ordered pair corresponds to point U?

34. Which ordered pair corresponds to point S?

35. Which point is located in Quadrant II?

Solve the equation for y.

36. $2x - y = 3$ 37. $3x + 2y = -4$ 38. $-2x = 6y + 3$

39. $0 = 7x - y + 12$ 40. $-2y + x = 4y - 6$

Solve the inequality. Graph the solution.

41. $p + 6 > 9$ 42. $3x - 4 < 2$

43. $-4m + 6 \leq 22$ 44. $5x + 1 \leq 3x - 9$

Answers

26. _____
27. _____
28. _____
29. _____
30. _____
31. _____
32. _____
33. _____
34. _____
35. _____
36. _____
37. _____
38. _____
39. _____
40. _____
41. ___See left.___
42. _____
    ___See left.___
43. _____
    ___See left.___
44. _____
    ___See left.___

Name _____ Date _____

Skills Test: Prerequisite Skills Test (continued)

Graph the equation.

45. $y - 2 = 2x$
46. $2y + x = 8$
47. $2x - 3y = 6$

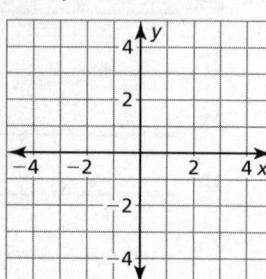
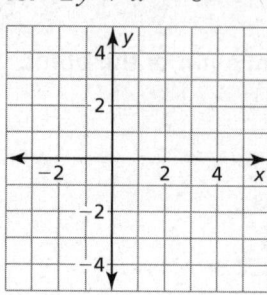
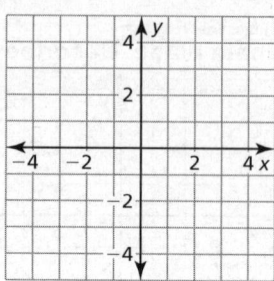

Evaluate the expression.

48. $14 \div 7 - 2^2 + (-3) \cdot 2 - 1$
49. $-4 - (3 + 6^2) \div 13 - 1^2 \cdot (-12)$

Find the square root(s).

50. $\sqrt{25}$
51. $-\sqrt{81}$
52. $\pm\sqrt{9}$
53. $-\sqrt{144}$

Write an equation for the *n*th term of the arithmetic sequence.

54. $3, 6, 9, 12, \ldots$
55. $7, 0, -7, -14, \ldots$
56. $2, 13, 24, 35, \ldots$

Simplify the expression.

57. $7x - 1 + 2x$
58. $3m + 2 - 6m + 8 - 1$
59. $-4(2y - 1) + 3y - 7$
60. $3(d + 3) - (2d - 1) + 11d + 8$

Evaluate the expression when $x = -3$.

61. $3x^2 - 6$
62. $2x^2 - 6x + 1$
63. $-x^2 - 5x - 1$
64. $x^2 + 3x + 8$
65. $-2x^2 + 4x + 3$
66. $-3x^2 - 6 - x$

Solve the system of linear equations by graphing.

67. $y = 2x + 1$
    $y = -2x - 3$
68. $y = -\frac{1}{2}x + 1$
    $y = x + 1$
69. $y = \frac{2}{3}x - 4$
    $y = -\frac{4}{3}x + 2$

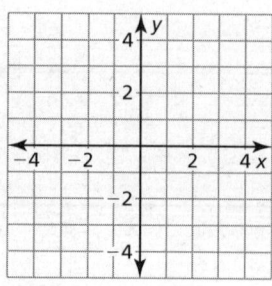
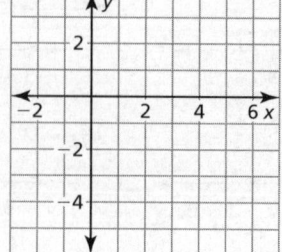

Answers

45. See left.
46. See left.
47. See left.
48. _____
49. _____
50. _____
51. _____
52. _____
53. _____
54. _____
55. _____
56. _____
57. _____
58. _____
59. _____
60. _____
61. _____
62. _____
63. _____
64. _____
65. _____
66. _____
67. _____
68. _____
69. _____

Name_____ Date_____

Skills Test: Prerequisite Skills Test (continued)

Find the greatest common factor.

70. 45, 9   **71.** 64, 48   **72.** 25, 10   **73.** 29, 12

Factor the trinomial.

74. $x^2 + 4x + 4$ **75.** $x^2 - 10x + 25$ **76.** $x^2 + 8x + 16$

77. $x^2 - 26x + 169$ **78.** $x^2 + 6x + 9$ **79.** $x^2 - 22x + 121$

Evaluate the expression.

80. $3\sqrt{9} - 6$ **81.** $\dfrac{\sqrt{25}}{15} - 7$

82. $2\left(\dfrac{\sqrt{16}}{8} + 6\right)$ **83.** $-3(9 - \sqrt{100})$

Graph f and g. Describe the transformations from the graph of f to the graph of g.

84. $f(x) = x;\ g(x) = -x - 1$ **85.** $f(x) = x;\ g(x) = 2x + 3$

86. The table shows the number of hours people spend on their cell phones daily. Display the data in a histogram.

Time spent	Frequency
0–1	3
2–3	6
4–5	10
6–7	1

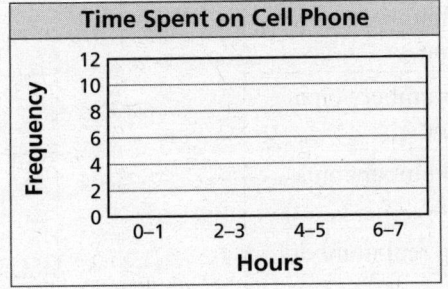

Answers

70. _____
71. _____
72. _____
73. _____
74. _____
75. _____
76. _____
77. _____
78. _____
79. _____
80. _____
81. _____
82. _____
83. _____
84. __See left.__
_____
_____
85. __See left.__
_____
_____
86. __See left.__

Prerequisite Skills Test Item Analysis

Item Number	Skills
1	adding and subtracting integers
2	adding and subtracting integers
3	adding and subtracting integers
4	adding and subtracting integers
5	adding and subtracting integers
6	adding and subtracting integers
7	multiplying and dividing integers
8	multiplying and dividing integers
9	multiplying and dividing integers
10	multiplying and dividing integers
11	multiplying and dividing integers
12	multiplying and dividing integers
13	specifying units of measure
14	specifying units of measure
15	graphing numbers on a number line
16	graphing numbers on a number line
17	graphing numbers on a number line
18	graphing numbers on a number line
19	comparing real numbers
20	comparing real numbers
21	comparing real numbers
22	comparing real numbers

Item Number	Skills
23	evaluating expressions
24	evaluating expressions
25	evaluating expressions
26	evaluating expressions
27	evaluating expressions
28	evaluating expressions
29	plotting points
30	plotting points
31	plotting points
32	plotting points
33	using a graph
34	using a graph
35	using a graph
36	rewriting equations
37	rewriting equations
38	rewriting equations
39	rewriting equations
40	rewriting equations
41	solving/graphing inequalities
42	solving/graphing inequalities
43	solving/graphing inequalities
44	solving/graphing inequalities

Prerequisite Skills Test Item Analysis (continued)

Item Number	Skills
45	graphing linear functions
46	graphing linear functions
47	graphing linear functions
48	using order of operations
49	using order of operations
50	finding square roots
51	finding square roots
52	finding square roots
53	finding square roots
54	writing equations for arithmetic sequences
55	writing equations for arithmetic sequences
56	writing equations for arithmetic sequences
57	simplifying algebraic expressions
58	simplifying algebraic expressions
59	simplifying algebraic expressions
60	simplifying algebraic expressions
61	evaluating expressions
62	evaluating expressions
63	evaluating expressions
64	evaluating expressions
65	evaluating expressions

Item Number	Skills
66	evaluating expressions
67	solving systems of linear equations by graphing
68	solving systems of linear equations by graphing
69	solving systems of linear equations by graphing
70	finding the greatest common factor
71	finding the greatest common factor
72	finding the greatest common factor
73	finding the greatest common factor
74	factoring perfect square trinomials
75	factoring perfect square trinomials
76	factoring perfect square trinomials
77	factoring perfect square trinomials
78	factoring perfect square trinomials
79	factoring perfect square trinomials
80	evaluating expressions involving square roots
81	evaluating expressions involving square roots
82	evaluating expressions involving square roots
83	evaluating expressions involving square roots
84	transforming linear functions
85	transforming linear functions
86	displaying data

Name _____ Date _____

Pre-Course Test

Solve the equation. Check your solution.

1. $x - 12 = 9$
2. $2x + 7 = -5 + x$

Solve the equation.

3. $8|2 - 9p| - 2 = 14$
4. $-3(-6x + 6) + 6(4x - 3) = -78$

Describe the values of *c* for which the equation has no solution.

5. $5x + 4 = 5x - c$
6. $|2x + 6| = c$

Write the sentence as an inequality.

7. A number *m* increased by 12 is less than 48.

8. The product of *x* and 10 is greater than or equal to 23.

Solve the inequality. Graph the solution.

9. $-6x - (-7x - 1) \le 6$
10. $2|2 - x| + 4 \le 16$

Write and graph a compound inequality that represents the numbers that are *not* solutions of the inequality represented by the graph shown.

11.

12.

Determine whether the relation is a function. If the relation is a function, determine whether the function is *linear* or *nonlinear*.

13.
x	1	2	3	4	5
y	1	4	9	16	25

14. $y = 3x + 1$

Write an equation in slope-intercept form of the line with the given characteristics.

15. through: $(3, -3)$ and $(-3, 4)$

16. through: $(-3, 5)$, parallel to $y = 2x + 2$

Answers

1. _____
2. _____
3. _____
4. _____
5. _____
6. _____
7. _____
8. _____
9. See left.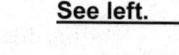
10. _____
    See left.
11. _____
    See left.
12. _____
    See left.
13. _____
    _____
14. _____
    _____
15. _____
16. _____

Pre-Course Test (continued)

Write an equation in point-slope form of the line with the given characteristics.

17. through: $(5, -3)$, slope $= -\dfrac{1}{5}$

18. through: $(-4, 2)$, perpendicular to $y = \dfrac{2}{3}x - 2$

Graph the equation and identify the intercept(s). If the equation is linear, find the slope of the line.

19. $3x - 4y = 12$

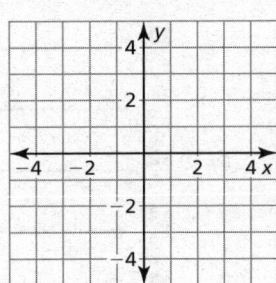

20. $|x + 2| - 3 = y$

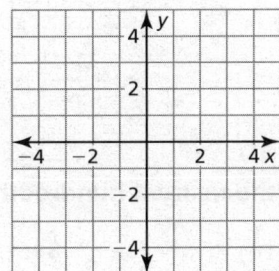

Graph f and g. Describe the transformations from the graph of f to the graph of g.

21. $f(x) = x$; $g(x) = \dfrac{1}{2}x - 1$

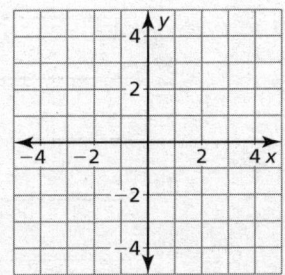

22. $f(x) = |x|$; $g(x) = 2|x + 3|$

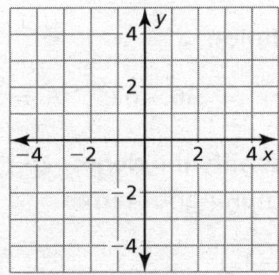

Graph the function. Describe the domain and range.

23. $f(x) = \begin{cases} -x + 4, & \text{if } x < -2 \\ \dfrac{1}{2}x - 3, & \text{if } x \geq -2 \end{cases}$

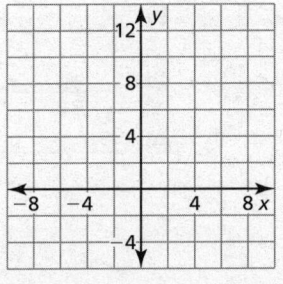

24. $y = \begin{cases} -\dfrac{3}{2}x - 2, & \text{if } x \leq 0 \\ -2, & \text{if } 0 < x < 3 \\ x - 5, & \text{if } x \geq 3 \end{cases}$

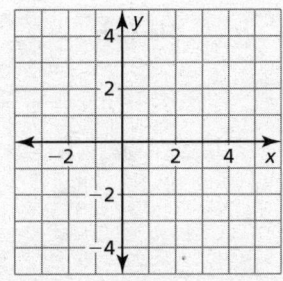

Answers

17. _____

18. _____

19. **See left.**

20. **See left.**

21. **See left.**

22. **See left.**

23. **See left.**

24. **See left.**

Name _____ Date _____

Pre-Course Test (continued)

Solve the system of linear equations using any method.

25. $-6x - 7y = 12$
    $6x + 10y = 6$

26. $7x + 2y = 29$
    $\frac{1}{2}y = -1 + x$

Graph the system of linear inequalities.

27. $y \geq 2x - 3$
    $x + y < 3$

28. $x - 2y < -6$
    $x + 2y < 2$

Evaluate the expression. Round to the nearest hundredth, if necessary.

29. $\sqrt[4]{49}$

30. $-(216)^{1/6}$

31. $81^{5/4}$

Simplify the expression. Write your answer using only positive exponents.

32. $m^5 \cdot m^{-4} \cdot m^0$

33. $\dfrac{ab^{-2}}{b^{-4}b^3}$

34. $\left(-\dfrac{3x^3}{2y}\right)^{-2}$

Solve the equation. Check your solution.

35. $3^{-2b} = \dfrac{1}{27}$

36. $625^{2-2p} = 25^{2p-1}$

Find the sum or difference. Then identify the degree of the sum or difference and classify the polynomial by the number of terms.

37. $(8 + 2x^2 + 7x^3) - (x^2 + 2 - 5x^3)$

38. $(5x^2 - 6x^4) + (7x^2 + 3x^4)$

Find the product.

39. $(m + 6)(m - 4)$

40. $(3a - 6)(4a + 6)$

41. $(x - 7)(x + 7)$

Factor the polynomial completely.

42. $x^3 - 2x^2 - 2x + 4$

43. $n^2 - 4n + 3$

44. $-2v^2 - 23v - 45$

Solve the equation.

45. $x(x - 4)(2x + 1) = 0$

46. $g^2 - 3g = 10$

Answers

25. _____
26. _____
27. _See left._
28. _See left._
29. _____
30. _____
31. _____
32. _____
33. _____
34. _____
35. _____
36. _____
37. _____
38. _____
39. _____
40. _____
41. _____
42. _____
43. _____
44. _____
45. _____
46. _____

Name_____ Date_____

Pre-Course Test (continued)

Solve the equation using any method.

47. $2x^2 - 98 = 0$ **48.** $x^2 - 10x + 2 = 0$ **49.** $2x^2 + 3x - 1 = 0$

Graph the function. Compare the graph to the graph of $f(x) = x^2$.

50. $g(x) = -\frac{1}{2}x^2 + 4$ **51.** $h(x) = 2(x - 1)^2 - 3$

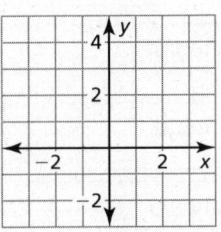 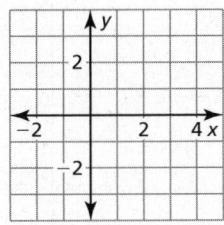

Find the inverse of the function.

52. $f(x) = -2x - 5$ **53.** $f(x) = -2(x - 4)^2 + 1,\ x \geq 4$

Graph the function f. Describe the domain and range. Compare the graph of f to the graph of g.

54. $f(x) = -\sqrt{x} + 3;\ g(x) = \sqrt{x}$ **55.** $f(x) = \sqrt[3]{x + 2};\ g(x) = \sqrt[3]{x}$

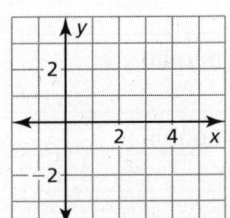 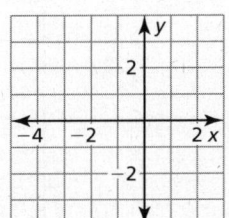

Solve the equation. Check your solutions.

56. $10 + \sqrt{\dfrac{m}{6}} = 12$ **57.** $n + 2 = \sqrt{6n + 3}$ **58.** $\sqrt{p + 5} = \sqrt{2p + 1}$

Find the mean, median, mode, and range of the data set. Round to the nearest tenth, if necessary.

59.

Student Test Scores			
84	76	92	99
91	65	88	78
69	88	75	87

60. 1, 4, 4, 11, 6, 4, 2, 9, 2

Answers

47. _____
48. _____
49. _____
50. _See left._
51. _See left._
52. _____
53. _____
54. _See left._
55. _See left._
56. _____
57. _____
58. _____
59. _____
60. _____

Pre-Course Test Item Analysis

Item Number	Skills
1	solving equations
2	solving equations
3	solving equations
4	solving equations
5	number sense
6	number sense
7	writing inequalities
8	writing inequalities
9	solving inequalities
10	solving inequalities
11	writing and graphing inequalities
12	writing and graphing inequalities
13	understanding functions
14	understanding functions
15	writing linear equations
16	writing linear equations
17	writing linear equations
18	writing linear equations
19	graphing linear equations
20	graphing linear equations
21	comparing linear transformations
22	comparing linear transformations
23	graphing piecewise functions
24	graphing piecewise functions

Item Number	Skills
25	solving linear systems
26	solving linear systems
27	graphing linear inequalities
28	graphing linear inequalities
29	evaluating rational expressions
30	evaluating rational expressions
31	evaluating rational expressions
32	simplify exponents
33	simplify exponents
34	simplify exponents
35	solving exponential equations
36	solving exponential equations
37	simplify polynomial expressions
38	simplify polynomial expressions
39	simplify polynomial expressions
40	simplify polynomial expressions
41	simplify polynomial expressions
42	factoring
43	factoring
44	factoring
45	solving quadratics
46	solving quadratics
47	solving quadratics
48	solving quadratics
49	solving quadratics

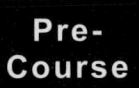

Pre-Course Test Item Analysis (continued)

Item Number	Skills
50	graphing quadratics
51	graphing quadratics
52	finding inverses
53	finding inverses
54	graphing radicals
55	graphing radicals

Item Number	Skills
56	solving radical equations
57	solving radical equations
58	solving radical equations
59	measures of center
60	measures of center

Name _____ Date _____

Chapter 1 Quiz
For use after Section 1.3

Solve the equation.

1. $x + 7 = 3$
2. $5.3 = z - 2.7$
3. $72 = -6r$
4. $\dfrac{1}{5}p = 15$
5. $4m - 6 = 14$
6. $8 = 11 - v$
7. $9 = -7w + 12w - 6$
8. $40 = -\dfrac{1}{3}(9x + 30) + 2$

Solve the equation. Determine whether the equation has *one solution*, *no solution*, or *infinitely many solutions*.

9. $6c + 3 = c - 12$
10. $4(2q - 3) = 8q - 12$
11. $-3(g - 4) = 2 - 3g$
12. $8(y - 1) - 3y = 6(2y - 6)$

13. To estimate your average monthly salary, divide your yearly salary by the number of months in a year. Write and solve an equation to determine your yearly salary when your average monthly salary is $4560.

14. You are a contractor and charge $45 for a site visit plus an additional $24 per hour for each hour you spend working at the site. Write and solve an equation to determine how many total hours you have to work to earn $810 working at two separate work sites.

15. You and a friend are both traveling from the Seattle area. You start 38 miles east of Seattle and travel east on Interstate 90 at 62 miles per hour. Your friend starts 20 miles south of Seattle and travels south on Interstate 5 at 65 miles per hour.

    a. Who would be farther from Seattle in two hours and by how much?

    b. How many hours will it take for you and your friend to be the same distance from Seattle?

Answers

1. _____
2. _____
3. _____
4. _____
5. _____
6. _____
7. _____
8. _____
9. _____
10. _____
11. _____
12. _____
13. _____
14. _____
15. a. _____

    b. _____

Chapter 1 Test A

Solve the equation. Justify each step.

1. $x + \dfrac{1}{2} = \dfrac{3}{4}$

2. $\dfrac{z}{4} = 12$

Solve the equation. Determine whether the equation has *one solution*, *no solution*, or *infinitely many solutions*.

3. $5n = -20$

4. $g + 5 = 17$

5. $13 + 3p + 10 = 23 + 3p$

6. $7 + 4y = 39$

7. $3 = t + 11.5 - t$

8. $4x + 8 + 6x - 5 = 33$

9. $5(2c + 7) - 3c = 7(c + 5)$

10. $\dfrac{3}{2}b + 6 + \dfrac{1}{2}b = 15 + 2b$

Describe the values of *c* for which the equation has no solution.

11. $2x - 6 = 2x - c$

12. $|x + 8| = c$

Find the value of the variable. Then find the angle measures of the polygon.

13.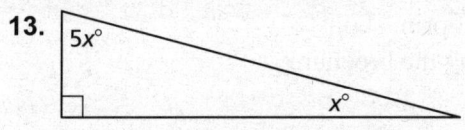

Sum of angle measures: 180°

14.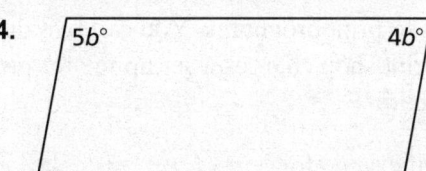

Sum of angle measures: 360°

Solve the equation.

15. $7n + 3 = 2n + 23$

16. $\dfrac{1}{2}(6x + 4) = 5(2x - 8)$

17. $\dfrac{3}{2}(d + 12) = \dfrac{1}{2}(2d - 6)$

18. $|b - 12| = 15$

19. $|2r + 5| = 3r$

20. $|2k + 6| = |k|$

Answers

1. _____
2. _____
3. _____
4. _____
5. _____
6. _____
7. _____
8. _____
9. _____
10. _____
11. _____
12. _____
13. _____
14. _____
15. _____
16. _____
17. _____
18. _____
19. _____
20. _____

Name _____ Date _____

Chapter 1 Test A (continued)

Solve the literal equation for y.

21. $5x + y = 2$

22. $2x + 5y = 3y + 8$

23. The formula for the volume of a cylinder is $V = \pi r^2 h$.

    a. Solve the formula for the height h.

    b. A cylinder has a volume of 628 cubic inches and a radius of 10 inches. What is the height of the cylinder rounded to the nearest inch?

24. The measures of two angles of a triangle are each four times the measure of the third angle. What is the measure of the third angle?

25. At a book fair, a tote bag costs $5 and books cost $3.50 each. You spend a total of $19 before taxes. How many books did you buy in addition to the tote bag?

26. For a school play, the maximum age for a youth ticket is 18 years old. The minimum age is 10 years old. Write an absolute value equation for which the two solutions are the minimum and maximum ages for a youth ticket.

27. Your business needs to print brochures. You call two different print shops about prices. Each print shop charges a set-up fee for preparing the brochure and a price per brochure.

    a. The total cost is the same for each company. How many brochures is your business printing?

    b. You decide to increase the number of brochures. From which company should you order?

	Brochure set-up fee	Price per brochure
Company A	$50	$1.50
Company B	$75	$1.00

Answers

21. _____

22. _____

23. a. _____

b. _____

24. _____

25. _____

26. _____

27. a. _____

b. _____

Name_____ Date_____

Chapter 1 Test B

Solve the equation. Justify each step.

1. $x + \dfrac{2}{3} = \dfrac{5}{6}$
2. $w - 8 = 12$

Solve the equation. Determine whether the equation has *one solution*, *no solution*, or *infinitely many solutions*.

3. $6m = -72$
4. $\dfrac{n}{3} = 15$
5. $5 + 2y = -13 + 2y$
6. $4h - 6 = 12$
7. $5 - k = 8 - k - 3$
8. $3x + 5 - 2x + 10 - x = 0$
9. $6(3 - d) + 2d = 24$
10. $\dfrac{1}{4}w + \dfrac{1}{2}w + 5 = 11$

Describe the value of *c* for which the equation is an identity.

11. $2(x + 5) = 2(x + 3) + c$
12. $|2x + 5| = |cx + 3 - 4x + 2|$

Find the value of the variable. Then find the angle measures of the polygon.

13.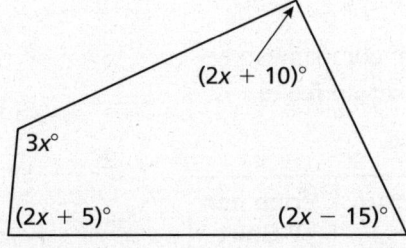

Sum of angle measures: 360°

14.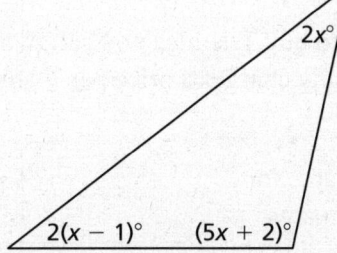

Sum of angle measures: 180°

Solve the equation.

15. $2n - 3 = 6n + 9$
16. $\dfrac{1}{2}(6x + 2) = 5(x + 3)$
17. $\dfrac{2}{3}(w + 12) = 3w - 6$
18. $|m + 8| = 12$
19. $|5y + 2| = 7y$
20. $|4k + 5| = |3k - 2|$

Answers

1. _____
   _____
2. _____
   _____
3. _____
4. _____
5. _____
6. _____
7. _____
8. _____
9. _____
10. _____
11. _____
12. _____
13. _____
    _____
14. _____
    _____
15. _____
16. _____
17. _____
18. _____
19. _____
20. _____

Chapter 1 Test B (continued)

Solve the literal equation for y.

21. $3x + 2y = 12$

22. $7x - 4y = 3y - 14$

23. The volume V of a cone is given by the formula $V = \frac{1}{3}\pi r^2 h$, where r is the radius of the base and h is the height.

   a. Solve the formula for height h.

   b. A cone has a volume of 120π cubic centimeters and a radius of 6 centimeters. What is the height of the cone?

24. A rectangular garden has a length that is five less than twice the width. The garden perimeter is 50 meters. What are the dimensions of the garden?

25. A necklace chain costs $15. Beads cost $2.75 each. You spend a total of $28.75 on a necklace and beads before tax. How many beads did you buy in addition to the necklace?

26. Consider the equation $\left|\frac{1}{4}x + 12\right| = \frac{x}{2}$. Without calculating, how do you know $x = -16$ is an extraneous solution?

27. Your soccer team wants to buy T-shirts. You call two different companies about prices. Each company charges a price per T-shirt and a set-up fee to create the team logo.

   a. The total cost is the same for each company. How many T-shirts is the team buying?

   b. A few players decide not to get T-shirts. Which company has a lower cost?

	Logo set-up fee	Price per T-Shirt
Company A	$50	$15
Company B	$95	$12

Answers

21. _____

22. _____

23. a. _____

    b. _____

24. _____

    _____

25. _____

26. _____

    _____

27. a. _____

    b. _____

Chapter 1 Alternative Assessment

1. Which of the following equations have only one solution? Which have two solutions? Which have no solution? Which have infinitely many solutions?

   a. $\frac{3}{5}x - 8 = 1$

   b. $5x - 2(x - 2) = 7x + 4 - 4x$

   c. $|3x + 2| - 2 = 6$

   d. $3|x - 2| + 4 = 2$

   e. $\frac{1}{2}(6x - 3) - x = 2x + 1$

   f. $|8x - 5| = |3x + 5|$

2. A doormat is in the shape of a trapezoid. The area A of the doormat is represented by the formula $A = \frac{1}{2}h(b_1 + b_2)$.

   a. Solve the formula for h.

   b. Show that solving the formula for b_1 by first multiplying both sides by 2 and then dividing both sides by h leads to $b_1 = \frac{2A}{h} - b_2$.

   c. Show that solving the formula for b_1 by first multiplying both sides by 2 and then using the Distributive Property to distribute the h leads to $b_1 = \frac{2A - b_2 h}{h}$.

   d. Show that the final formula in part (b) is equivalent to the final formula in part (c) by showing the steps to transform one formula to the other one.

   e. Explain how you could solve the original formula for b_2.

Name _____ Date _____

Chapter 1 — Alternative Assessment Rubric

Score	Conceptual Understanding	Mathematical Skills	Work Habits
4	Shows complete understanding of: • solving linear equations and the types of solutions • rewriting equations to solve for one variable in terms of the other variable(s)	Correctly answers all questions in Exercise 1 Correctly follows all directions in the steps to rewrite the equation in Exercise 2	Answers all parts of both problems Answers are explained thoroughly with mathematical terminology. Work is very neat and well organized.
3	Shows nearly complete understanding of: • solving linear equations and the types of solutions • rewriting equations to solve for one variable in terms of the other variable(s)	Correctly answers most questions in Exercise 1 Correctly follows most directions in the steps to rewrite the equation in Exercise 2	Answers several parts of both problems Answers are explained with mathematical terminology. Work is neat and well organized.
2	Shows some understanding of: • solving linear equations and the types of solutions • rewriting equations to solve for one variable in terms of the other variable(s)	Correctly answers some questions in Exercise 1 Incorrectly follows directions in the steps to rewrite the equation in Exercise 2	Answers some parts of both problems Answers are poorly or incorrectly explained. Work is not very neat or organized.
1	Shows little understanding of: • solving linear equations and the types of solutions • rewriting equations to solve for one variable in terms of the other variable(s)	Does not answer Exercise 1 Does not correctly answer Exercise 2	Does not attempt any part of either problem No explanation is included. Work is sloppy and disorganized.

Name_____ Date_____

 **Performance Task**

Magic of Mathematics

Instructional Overview	
Launch Question	Have you ever watched a magician perform a number trick? You can use algebra to explain how these types of tricks work.
Summary	There are two algebra magic problems through which students should work with a partner, using algebra to prove why each answer works. Then students create their own magic problem by working backward, and then testing it with their peers.
Teacher Notes	To introduce the task, work out the following problem with your class. Ask your students to think of a number, any positive integer (small is best because it allows them to do computations mentally). Step 1: Think of any number. Step 2: Square it. Step 3: Add the result to your original number. Step 4: Divide by your original number. Step 5: Add 24. Step 6: Subtract your original number. Step 7: Divide by 5. Your number is 5. To help students get started on their task, run through the algebra of this magic number trick with them. Step 1: x Step 2: x^2 Step 3: $x^2 + x$ Step 4: $\dfrac{x^2 + x}{x} = x + 1$ Step 5: $(x + 1) + 24 = x + 25$ Step 6: $(x + 25) - x = 25$ Step 7: $\dfrac{25}{5} = 5$ You can have students work with partners on the problems, or you could run through the magic problems together with the students. Then have them work individually on the task.

Name _____ Date _____

 **Performance Task** (continued)

Instructional Overview (continued)	
Supplies	Copies of the task
Mathematical Discourse	Where have you seen problems of this type? How does working through the algebra change the magic of the problem?
Writing/Discussion Prompts	1. Describe a time you saw a magic trick that amazed you. Do you think it could be explained using mathematics? 2. Write a simple magic number trick and explain it with algebra. 3. Why do you think people are amazed by problems of this type?

Magic of Mathematics

Curriculum Content	
Content Objectives	• Create equations in one variable to focus on a quantity of interest. • Justify each step in solving a simple equation. • Solve linear equations in one variable.
Mathematical Practices	• Use problem-solving skills to determine a way to show how the number trick works. • Use algebra to model and explain a number trick.

Chapter 1 Performance Task (continued)

Rubric

Magic of Mathematics	Points	
1. Step 1: x Step 2: $2x$ Step 3: $2x + 5$ Step 4: $2(2x + 5)$ Step 5: $2(2x + 5) - 6$ Step 6: $\dfrac{2(2x + 5) - 6}{4}$ Step 7: $\dfrac{2(2x + 5) - 6}{4} - x$ Simplify: $\dfrac{2(2x + 5) - 6}{4} - x = \dfrac{4x + 10 - 6}{4} - x$ $= \dfrac{4x + 4}{4} - x = (x + 1) - x = 1$	3 2 1	The algebra is correct. Most of the steps are correct, but there are some errors in a few steps. Only a few steps are correct.
2. Step 1: x Step 2: $x + 3$ Step 3: $2(x + 3)$ Step 4: $2(x + 3) - 9$ Step 5: $5[2(x + 3) - 9]$ Step 6: $5[2(x + 3) - 9] + 15$ Step 7: $\dfrac{5[2(x + 3) - 9] + 15}{10}$ Simplify: $\dfrac{5[2(x + 3) - 9] + 15}{10} = \dfrac{5(2x + 6 - 9) + 15}{10}$ $= \dfrac{5(2x - 3) + 15}{10}$ $= \dfrac{10x - 15 + 15}{10} = \dfrac{10x}{10} = x$ Magic Problem 1 always returns the same number (1), whereas Magic Problem 2 always returns the originally chosen number.	4 3 2 1	The algebra and explanation are correct. The algebra is correct. Most of the steps are correct, but there are some errors in a few steps. Only a few steps are correct.

Name _____ Date _____

 **Performance Task** (continued)

Rubric (continued)

Magic of Mathematics	Points	
3. This should be a magic trick similar to the previous ones.	5	The trick works, has at least three steps, and the algebra is correct.
	3	Most of the steps are correct, but there are some errors in a few steps, or there is not algebraic explanation.
	1	Only a few steps are correct, and there is no supporting algebra.
Mathematics Practice: Students use problem solving to complete the task. The student understands that algebra is being used to model the problem.	3	The student demonstrates problem-solving skills, perseverance, and the ability to model the problem using algebra. Award partial credit as needed.
	Total Points	15 points

Performance Task (continued)

Magic of Mathematics

Have you ever watched a magician perform a number trick? You can use algebra to explain how these types of tricks work.

Work through problems 1 and 2 with a partner. Take turns having one of you read the steps aloud as the other calculates and records the results. Then repeat the steps with a variable in place of a specific number. Work together using algebra to explain why the magic trick works.

1. Step 1: Think of any number.
   Step 2: Double the number.
   Step 3: Add 5.
   Step 4: Double your new number.
   Step 5: Subtract 6.
   Step 6: Divide by 4.
   Step 7: Subtract your original number.
   Did you get 1? Use algebra to show why this works.

2. Step 1: Select a two-digit number less than 100.
   Step 2: Add 3 to your number.
   Step 3: Double the result.
   Step 4: Subtract 9.
   Step 5: Multiply by 5.
   Step 6: Add 15.
   Step 7: Divide by 10.
   Did you get your original number? Use algebra to show why this works. How is this problem different from the previous problem?

3. Create your own magic number trick and use algebra to explain why it works.

Name _____ Date _____

Chapter 2 Quiz
For use after Section 2.4

Write the sentence as an inequality.

1. A number q plus 8 is less than or equal to 15.

2. The number 20 is no less than the difference of 4 times a number c and 8.

Write an inequality that represents the graph.

3.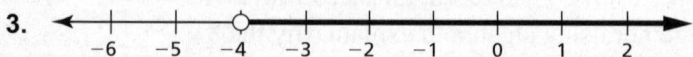

4.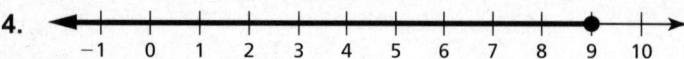

Solve the inequality.

5. $6 + t \leq 13$

6. $m - (-4) < 8$

7. $6s \geq 24$

8. $9 > \dfrac{p}{-3}$

Solve the inequality.

9. $5y - 4 \geq 26$

10. $-18 < 4t + 6$

11. $4(4b - 6) \leq 8(2b + 4)$

12. $3(2x + 7) \leq 6(x - 4)$

13. Two requirements to be employed within a certain company are that you must be at least 18 years of age and have had more than one job-related work experience.

    a. Write and graph two inequalities that represent the requirements.

    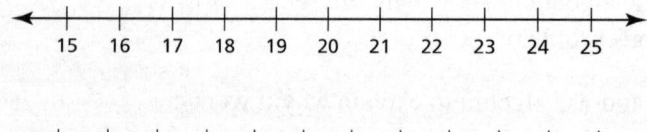

    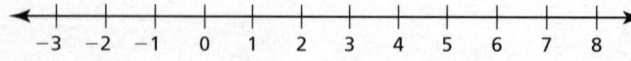

    b. You are 19 years old and have had one job-related work experience. Do you satisfy the requirements to apply?

14. The *American Heart Association* recommends that you consume less than 1500 milligrams of sodium per day. Today you have already taken in 400 milligrams of sodium. Write and solve an inequality that represents the additional amount of sodium you can consume.

Answers

1. _____
2. _____
3. _____
4. _____
5. _____
6. _____
7. _____
8. _____
9. _____
10. _____
11. _____
12. _____
13. a. _____

    See left.

    b. _____

14. _____

    _____

Chapter 2 Test A

Write the sentence as an inequality.

1. The sum of twice a number *n* and 8 is at most 25.

2. The temperature *t* is at least 75°F.

3. The cost of a ticket *t* will be no more than $26.

Write an inequality that represents the graph.

4.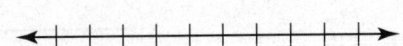

5.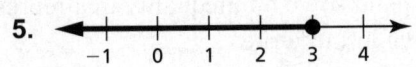

Solve the inequality. Graph the solution.

6. $-9 < m - 6$

7. $-3z \geq 6 + 3z$

Solve the inequality.

8. $m \geq 5m - 4$

9. $\dfrac{x}{4} + 6 \leq x + 8$

10. $\dfrac{1}{2}h + 2 \geq \dfrac{1}{2}(h + 8)$

11. $4k - (3 + 3k) > 2$

12. $4n + 3 < 6n + 8 - 2n$

13. $10 - 2(3x - 1) > 6x + 10$

Solve the inequality. Graph the solution.

14. $-3y > 9$ or $2y - 6 > 2$

15. $-1 < c + 2 < 3$

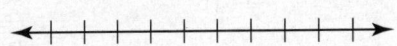

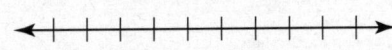

Solve the inequality.

16. $2a + 1 < 11$ or $a < 3a - 12$

17. $32 > 16 - 4g > 12$

18. $|2x - 6| < 0$

19. $|7 - 2y| - 8 \geq -3$

Answers

1. _____
2. _____
3. _____
4. _____
5. _____
6. _See left._
7. _____
   See left.
8. _____
9. _____
10. _____
11. _____
12. _____
13. _____
14. _____
    See left.
15. _____
    See left.
16. _____
17. _____
18. _____
19. _____

Chapter 2 Test A (continued)

Write and graph a compound inequality that represents the numbers that are not solutions of the inequality represented by the graph shown.

20. 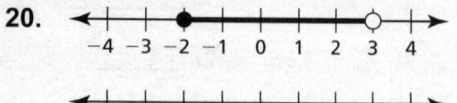   21.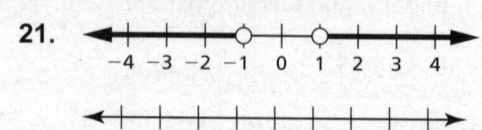

Answers

20. _____
    See left.

21. _____
    See left.

22. You need to write an essay that has at least 500 words. You have written 285 words so far. Write and solve an inequality that represents the number of words w that you have left to write.

22. _____
    _____

23. You need at least 30 cubic feet of sand to fill a sand box. Each bag contains 2.5 cubic feet of sand. Write and solve an inequality that represents the number of bags b that you need to buy.

23. _____
    _____

24. You are planning a school carnival. The equipment costs $180 to rent. You plan to charge $4.00 per ticket. You would like to have a profit of at least $500. Write and solve an inequality that represents the number of tickets t that you need to sell.

24. _____
    _____

25. You want to purchase a calculator for at most $115. You have saved $30 so far. You earn $7.50 per hour at your job. Write and solve an inequality that represents the number of hours h that you need to work.

25. _____
    _____

Chapter 2 Test B

Write the sentence as an inequality.

1. The product of a number n and 2 is no less than 14.

2. The speed s on a highway is at most 60 miles per hour.

3. The length r of a rope should be at least 28 inches.

Write an inequality that represents the graph.

4. (number line from −1 to 7, open circle at 5, shaded left)

5. (number line from −2 to 6, open circles at −1 and 5, shaded between)

Solve the inequality. Graph the solution.

6. $x + 5 \leq -2$

7. $4q > -28$

Solve the inequality.

8. $2k > 2k + 4$

9. $4p < 6p + 12$

10. $2.5w - 5 < 2w + 5$

11. $5(p - 1) > 6p - 7$

12. $5n + 3 \geq 4 - (6 - 5n)$

13. $5 - 2x < 4 - 2x + 3$

Solve the inequality. Graph the solution.

14. $5 + 2y < 8$ or $5y > 3y + 7$

15. $7 < 12 + c < 13$

Solve the inequality.

16. $-3p + 1 \leq -11$ or $-0.5p > 12$

17. $6 < 4 - w \leq 2w - 2$

18. $|3x + 15| < 6$

19. $3 - |x + 8| \geq 5$

Answers

1. _____
2. _____
3. _____
4. _____
5. _____
6. ___See left.___
7. _____
   ___See left.___
8. _____
9. _____
10. _____
11. _____
12. _____
13. _____
14. _____
   ___See left.___
15. _____
   ___See left.___
16. _____
17. _____
18. _____
19. _____

Chapter 2 Test B (continued)

Write and graph a compound inequality that represents the numbers that are not solutions of the inequality represented by the graph shown.

20.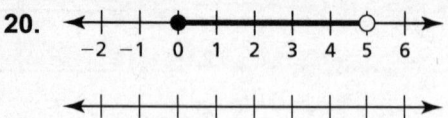

21.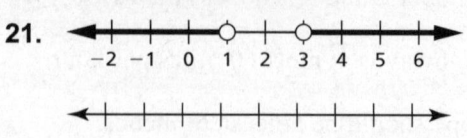

22. You need to earn at least $75. You earn $6.00 for each hour you work. Write and solve an inequality that represents the number of hours h that you need to work.

23. You need at least 150 cups of lemonade but less than 225 cups of lemonade for a picnic. Each batch of lemonade makes 25 cups of lemonade. Write and solve an inequality that represents the number of batches b you need to make.

24. You have a goal to practice the piano for an average of at least 50 minutes per day for one week. The first six days you practice a total of 245 minutes. Write and solve an inequality that represents the number of minutes m you need to practice on the seventh day.

25. The cost to rent a construction crane is $1500 per day plus $250 per hour of use. Write and solve an inequality that can be used to determine the maximum number of hours h the crane can be used if the rental cost for one day will not exceed $5000.

Answers

20. _____
    See left.
21. _____
    See left.
22. _____
    _____
23. _____
    _____
24. _____
    _____
25. _____
    _____

Chapter 2 Alternative Assessment

1. Which of the following inequalities have $x \geq 1$ as all or part of the solution? Which have no solution? Which have a solution set of all real numbers?

   a. $|2x + 5| + 10 \geq 3$

   b. $\frac{1}{2}(8x + 10) - 7 \geq \frac{1}{3}(9x + 6) + x$

   c. A number x plus sixteen is no less than 17.

   d. $-3 \leq 4x - 7 \leq 13$

   e. $|x - 5| + 4 \leq 1$

   f. $21 - 3x \leq 18$

   g. $9x + 6 < 12x - 3x + 6$

2. Your friend's dog-grooming business has a weekly profit that is modeled by the expression $3x - 5$.

   a. Your friend's weekly profit is more than $40 and at most $130. Write your friend's weekly profit as an inequality.

   b. Solve the inequality.

   c. Graph the inequality.

   d. The weekly profit of your cat-grooming business is modeled by the expression $2x - 15$. Your profit is more than $55 and at most $105. Write your profit as an inequality.

   e. Solve the inequality.

   f. Graph the inequality representing your friend's profit and the inequality representing your profit on the same number line.

   g. Use the graph to determine if it is possible for you and your friend to earn the same weekly profit. Explain.

   h. Determine the values of x for which you and your friend could earn the same weekly profit, if applicable.

Name _____ Date _____

Chapter 2 Alternative Assessment Rubric

Score	Conceptual Understanding	Mathematical Skills	Work Habits
4	Shows complete understanding of: • solving linear inequalities and the types of solutions • compound inequalities and their graphs	Writes and solves inequalities to answer all of Exercise 1 Writes, solves, and graphs compound inequalities Correctly determines the intersection of two inequalities	Answers all parts of both problems All inequalities and graphs are written or drawn carefully and systematically. Work is very neat and well organized.
3	Shows nearly complete understanding of: • solving linear inequalities and the types of solutions • compound inequalities and their graphs	Writes and solves inequalities to answer most of Exercise 1 Writes, solves, or graphs compound inequalities Partially determines the intersection of two inequalities	Answers several parts of both problems Most inequalities and graphs are written or drawn carefully and systematically. Work is neat and organized.
2	Shows some understanding of: • solving linear inequalities and the types of solutions • compound inequalities and their graphs	Writes and solves inequalities to answer some of Exercise 1 Attempts to write, solve, or graph compound inequalities Attempts to determine the intersection of two inequalities	Answers some parts of both problems Inequalities and graphs are written or drawn carelessly. Work is not very neat or organized.
1	Shows little understanding of: • solving linear inequalities and the types of solutions • compound inequalities and their graphs	Does not answer Exercise 1 Does not write, solve, or graph compound inequalities Does not determine the intersection of two inequalities	Does not attempt any part of either problem No inequalities or graphs are written or drawn. Work is sloppy and disorganized.

Name_____ Date_____

 **Performance Task**

Grading Calculations

Instructional Overview	
Launch Question	You are not doing as well as you had hoped in one of your classes. So, you want to figure out the minimum grade you need on the final exam to receive the semester grade that you want. Is it still possible to get an A? How would you explain your calculations to a classmate?
Summary	Two problems are provided with four scores and the students are asked to figure out if they can achieve an average of 90% or more when they take the final exam. All scores are percents, so make sure they do not plan to include a final score greater than 100. In the last part of the task, the students are asked to explain a question to a peer.
Teacher Notes	Make sure the students know how to calculate the mean. Also, remind them that we are using percent scores, so 100 is the maximum score.
Supplies	Copies of the task
Mathematical Discourse	Do you ever try to figure out what grade you need to receive the final grade you want? How would you change the calculations to see what grade you need to get a B? What are the different ways to approach this problem?
Writing/Discussion Prompts	1. How would the calculations change if you were allowed to throw out the lowest score? 2. What is the greatest range you could have with five of your test scores and maintain a B average? 3. Why is it that students have trouble calculating their grades?

Curriculum Content	
Content Objectives	• Create inequalities in one variable. • Solve linear inequalities in one variable.
Mathematical Practices	• Construct viable arguments and use reasoning to explain a question to others. • Look beyond the algorithmic problem when an average is calculated to understand the components of calculating averages.

Name _____ Date _____

 **Performance Task** (continued)

Grading Calculations
Rubric

Grading Calculations	Points	
1a: The sum of the first four scores is 352. To have a 90%, you would need 450 points so that $\frac{450}{5} = 90$. Because $450 - 352 = 98$, it is possible, but difficult.	3	Answer is correct and steps are shown, some explanation is included
	2	Most of the steps are correct with minor errors in computation, attempted explanation
	1	A few steps are correct; The learner demonstrates some knowledge of the concept.
1b: The sum of the first four scores is 345. To have a 90%, you would need 450 points so that $\frac{450}{5} = 90$. Because $450 - 345 = 105$, it is not possible. The highest percent this student can earn for the semester with a 100% on the final is 89%.	3	Answer is correct and steps are shown, some explanation is included
	2	Most of the steps are correct with minor errors in computation, attempted explanation
	1	A few steps are correct; The learner demonstrates some knowledge of the concept.
2: The lowest possible score is 50. If the first 4 exams were the maximum of 100 each, the student would need 50 additional points on the final exam to reach a total of 450 points.	3	A reasonable explanation that supports the fact that an A can be earned without all As
	1	The answer does not demonstrate an understanding of the task.
Mathematics Practice: Look for and make use of structure. When an average is calculated, the students need to look beyond the algorithmic and use alternative approaches.	3	The student demonstrates an understanding of how averages are calculated and is able to make the needed modifications.
Total Points	12 points	

Name_____ Date_____

 **Performance Task** (continued)

Grading Calculations

You are not doing as well as you had hoped in one of your classes. So, you want to figure out the minimum grade you need on the final exam to receive the semester grade that you want. Is it still possible to get an A? How would you explain your calculations to a classmate?

1. The grading scale states that you must have a 90% to earn an A for the semester. There are five tests each semester (four exams and one final). Given the percentage scores on the first four exams, determine if it is possible for you to get an A with your score on the final. Remember, the maximum possible score is 100%.

    a. Exam 1 = 85, Exam 2 = 80, Exam 3 = 94, and Exam 4 = 93

    Can you earn an A? If so, what score would you need on the final exam? If not, what is the highest percent you can get for your semester grade? Explain.

    b. Exam 1 = 82, Exam 2 = 84, Exam 3 = 83, and Exam 4 = 96

    Can you earn an A? If so, what score would you need on the final exam? If not, what is the highest percent you can get for your semester grade? Explain.

2. What is the lowest single score (percent) you could get on one of the five exams and still receive an A? How did you calculate your score? Explain it to your friend who believes the only way to receive an A is by earning an A on every exam. Is she right? Why or why not?

Name _____ Date _____

Chapter 3 Quiz
For use after Section 3.3

Determine whether the relation is a function. Explain.

1.
Input, x	9	7	5	3	1
Output, y	1	2	2	3	4

2. $(5, 4), (3, 10), (1, 16), (3, 8), (-1, 6)$

Answers

1. _____

Find the domain and range of the function represented by the graph.

2. _____

3. 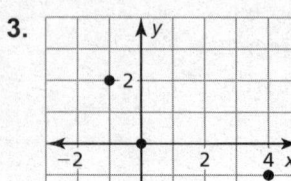   4. 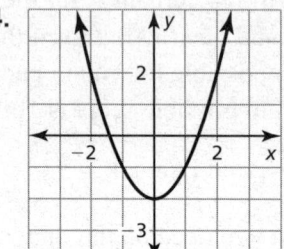   5.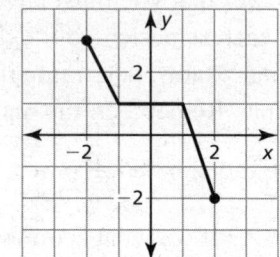

3. _____

4. _____

Determine whether the graph, table, or equation represents a *linear* or *nonlinear* function. Explain.

5. _____

6. 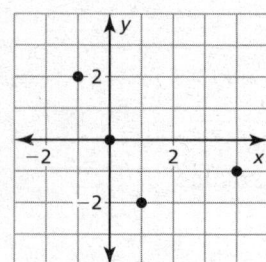   7.
x	y
−3	4
0	6
3	8

8. $y = x(x + 3)$

6. _____

7. _____

Graph the linear function.

9. $g(x) = x + 7$ 10. $p(x) = -2x - 4$ 11. $m(x) = \dfrac{3}{4}x$

8. _____

9. __See left.__

10. __See left.__

11. __See left.__

12. _____

12. The function $c = 15 + 9h$ represents the amount c (in dollars) it will cost you for a one-time lawn care service of h hours. Identify the independent and dependent variables. Is the domain discrete or continuous? Explain.

Name_____ Date_____

Chapter 3 Test A

Determine whether the relation is a function. If the relation is a function, determine whether the function is *linear* or *nonlinear*.

1.
x	1	3	5	3
y	4	2	0	-2

2. $y = 3$

3. $2x - 5y = 10$

4. $\dfrac{5}{x} + y = -7$

Find the domain and range of the function represented by the graph. Determine whether the domain is *discrete* or *continuous*.

5.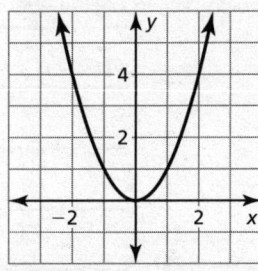

6.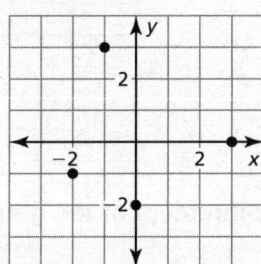

Evaluate the function when $x = -1, 0,$ and 4.

7. $g(x) = 3x^2 + 1$
8. $b(x) = -2x - 4$
9. $h(x) = |-x + 5|$

Find the value of x so that $f(x) = 3$.

10.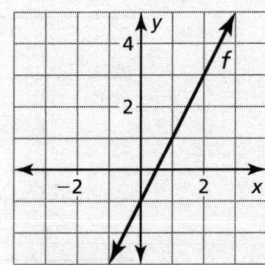

11.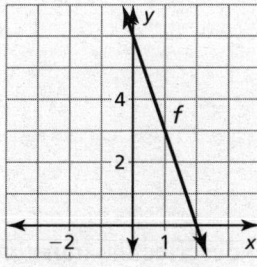

Find the x- and y-intercepts of the graph of the linear equation.

12. $2x + 3y = 6$
13. $-3x + 5y = -30$
14. $\dfrac{1}{2}x + y = -8$

Answers

1. _____
   _____
2. _____
   _____
3. _____
   _____
4. _____
   _____
5. _____
   _____
   _____
6. _____
   _____
7. _____
8. _____
9. _____
10. _____
11. _____
12. _____
    _____
13. _____
    _____
14. _____
    _____

Name _____ Date _____

Chapter 3 Test A (continued)

The points represented by the table lie on a line. Find the slope of the line.

15.
x	−5	−3	−1	1
y	7	4	1	−2

16.
x	2	2	2	2
y	−6	3	−7	1

Graph the linear equation.

17. $x - 3y = 6$

18. $y = -\frac{2}{3}x + 1$

Identify the slope, y-intercept, and x-intercept of the graph of the linear equation.

19. $5x + 3y = 15$ 20. $y = x - 3$ 21. $x = -4$

Use the graph of f and g to describe the transformation from the graph of f to the graph of g.

22.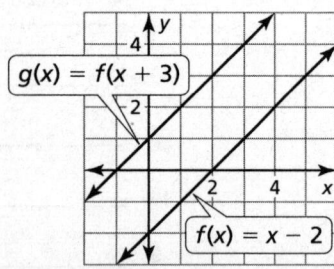

23. $f(x) = -x + 5;\ g(x) = 2f(x)$

24. Given $g(x) = -|x - 2| + 3$, (a) describe the transformation from the graph of $f(x) = |x|$ to the graph of g, and (b) graph g.

Answers

15. _____
16. _____
17. __See left.__
18. __See left.__
19. _____
_____
_____
20. _____
_____
_____
21. _____
_____
_____
22. _____
_____
_____
23. _____
_____
_____
_____
24. a. _____
_____
_____
b. __See left.__

Name_____ Date_____

Chapter 3 Test B

Determine whether the relation is a function. If the relation is a function, determine whether the function is *linear* or *nonlinear*.

1.
x	0	2	4	6
y	-8	-3	3	7

2.
x	0	1	2	3
y	-4	-2	0	2

3. $2y - 4 = 10$

4. $2xy = -8$

Find the domain and range of the function represented by the graph. Determine whether the domain is *discrete* or *continuous*.

5.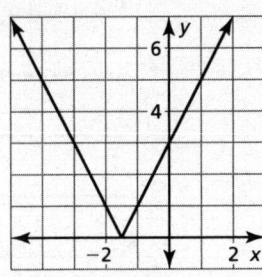

6.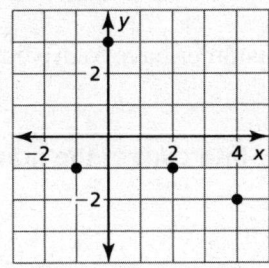

Evaluate the function when $x = -3, -2,$ and 1.

7. $g(x) = -x^2 - 7$

8. $h(x) = |-2x - 6|$

9. $f(x) = \frac{1}{2}x - 1$

Find the value of x so that the function has the given value.

10. $j(x) = 3 - x; j(x) = -5$

11. $t(x) = 2x - 4; t(x) = \frac{1}{2}$

12. $m(x) = -\frac{2}{3}x + 8; m(x) = 2$

13. $k(x) = \frac{3}{2}x - 1; k(x) = -4$

Find the x- and y-intercepts of the graph of the linear equation.

14. $2x - 3y = -10$

15. $2x + 5y = -8$

16. $-4 - x = 14 - 3y$

Graph the linear equation.

17. $2x - 3y = 9$

18. $-2y - 4 = 4$

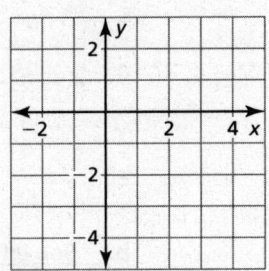

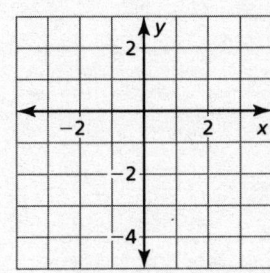

Answers

1. _____
2. _____
3. _____
4. _____
5. _____
   _____
   _____
   _____
6. _____
   _____
   _____
7. _____
8. _____
9. _____
10. _____
11. _____
12. _____
13. _____
14. _____
   _____
15. _____
   _____
16. _____
17. __See left.__
18. __See left.__

Chapter 3 Test B (continued)

The points represented by the table lie on a line. Find the slope of the line.

19.
x	1	−4	−3	2
y	3	3	3	3

20.
x	1	3	7	−1
y	−1	2	8	−4

21. The function $c = 100 + 0.30m$ represents the cost c (in dollars) of renting a car after driving m miles.

   a. Identify the independent and dependent variables.

   b. What would the cost be to rent the car and drive 100 miles?

   c. How many miles would a customer have to drive for the cost to be $149.50?

Identify the slope, y-intercept, and x-intercept of the graph of the linear equation.

22. $y = -x + 3$ 23. $4x - 6y = 14$ 24. $3y + 4 = -10$

Use the graphs of f and g to describe the transformation from the graph of f to the graph of g.

25. 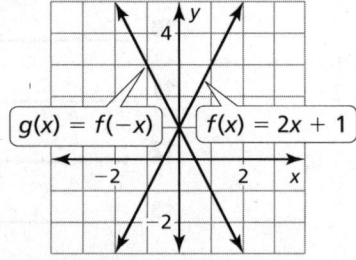 $g(x) = f(-x)$; $f(x) = 2x + 1$

26. $f(x) = 2x - 4$; $g(x) = \frac{1}{2}f(x)$

27. Given $g(x) = -2|x - 1| + 2$, (a) describe the transformation from the graph of $f(x) = |x|$ to the graph of g, and (b) graph g.

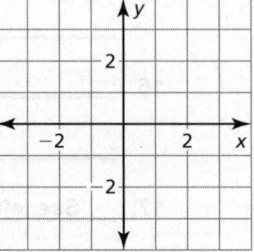

Answers

19. _____

20. _____

21. a. _____

    b. _____

    c. _____

22. _____

_____

_____

23. _____

_____

_____

24. _____

_____

_____

25. _____

_____

_____

26. _____

_____

_____

27. a. _____

_____

_____

   b.   See left.

Chapter 3 Alternative Assessment

1. The function $F = \frac{9}{5}C + 32$, $C \geq -273.15°$ relates the Fahrenheit and Celsius temperature scales.

   a. Identify the independent and dependent variables.

   b. Find the domain of the function. Is the domain discrete or continuous? Explain.

   c. What do the numbers $\frac{9}{5}$ and 32 represent?

   d. Graph the function.

   e. Evaluate the function when $C = 30°, -15°,$ and $-22°$.

   f. Find the value of C for which $F = 71°$.

   g. Find a negative value of C for which F is positive.

   h. Write the function using function notation.

2. Using $f(x) = 2x$, graph g. Describe the transformations from the graph of f to the graph of g.

   a. $g(x) = f(x) - 3$

   b. $g(x) = f(3x)$

   c. $g(x) = -f(x)$

   d. $g(x) = f(-x)$

   e. $g(x) = 3f(x)$

   f. $g(x) = f(x - 3)$

   g. $g(x) = 2f(x + 5)$

   h. $g(x) = -f(x) + 2$

   i. $g(x) = -4f(x + 1) + 2$

Name _____ Date _____

Chapter 3 Alternative Assessment Rubric

Score	Conceptual Understanding	Mathematical Skills	Work Habits
4	Shows complete understanding of: • functions and their characteristics • horizontal/vertical translations, reflections, and horizontal/vertical stretches/shrinks	Correctly answers all the function questions in Exercise 1 Correctly describes all the transformations from the graph of f to the graph of g	Answers all parts of both problems All equations and graphs are written or drawn carefully and systematically Work is very neat and well organized
3	Shows nearly complete understanding of: • functions and their characteristics • horizontal/vertical translations, reflections, and horizontal/vertical stretches/shrinks	Correctly answers most of the function questions in Exercise 1 Correctly describes most of the transformations from the graph of f to the graph of g	Answers several parts of both problems Most equations and graphs are written or drawn carefully and systematically Work is neat and organized
2	Shows some understanding of: • functions and their characteristics • horizontal/vertical translations, reflections, and horizontal/vertical stretches/shrinks	Correctly answers some of the function questions in Exercise 1 Correctly describes some of the transformations from the graph of f to the graph of g	Answers some parts of both problems Equations and graphs are written or drawn carelessly Work is not very neat or organized
1	Shows little understanding of: • functions and their characteristics • horizontal/vertical translations, reflections, and horizontal/vertical stretches/shrinks	Does not answer the function questions in Exercise 1 Does not describe the transformations from the graph of f to the graph of g	Does not attempt any part of either problem No equations or graphs are written or drawn Work is sloppy and disorganized

Name_____ Date_____

 **Performance Task**

The Cost of a T-Shirt

Instructional Overview	
Launch Question	You receive bids for making T-shirts for your class fundraiser from four companies. To present the pricing information, one company uses a table, one company uses a written description, one company uses an equation, and one company uses a graph. How will you compare the different representations and make the final choice?
Summary	The information will be provided to the students in different formats. The number of T-shirts needed is open-ended. When the four functions are created, the students will determine that different companies should be selected based on the number of T-shirts needed. The intervals for the best choice will be the same for all students, and the conclusion will be open as the students will use the information to make their own conclusions.
Teacher Notes	Part 1 can be completed individually, and then the students can work together to discuss and finalize the information. Part 2 can be completed in teams or individually after the groups complete their work. Encourage your students to create equations for each of the companies and then graph the information on the same coordinate plane to look for the points of intersection. The graph that is provided will provide clues to the coordinate plane that is to be graphed. Solution for Part 1: $y = 5x$, $20 < x < 40$; $y = 3.5x + 60$, $40 < x < 80$; $y = 2x + 180$, $x > 80$
Supplies	Graph paper, graphing calculator or computers (if used)
Mathematical Discourse	The students should personalize Part 2 by selecting how many T-shirts they will order, explaining why, and then selecting a company and supporting their choice from their data. After the group discussions, the students should complete their project by writing out their solution.

Curriculum Content	
Content Objectives	• Compare four functions that are each represented in a different format (table, written description, equation, and graph). • Create equations in two variables to represent relationships between quantities or graph equations on coordinate axes. • Relate the domain of a function to its graph.
Mathematical Practices	• Make sense of what is being asked before beginning to solve. • Reason abstractly and quantitatively to make the data available in one format. • Construct viable arguments and critique the reasoning of others to determine the best bid and the number of t-shirts to order.

Name _____ Date _____

Chapter 3 Performance Task (continued)

The Cost of a T-Shirt

Rubric

The Cost of a T-Shirt	Points	
Part 1: Represents all four proposals in the same format so that they can be compared Correct equations and/or graphs: $y_1 = 5x$ $y_2 = 3.5x + 60$ $y_3 = 2x + 180$ $y_4 = 2.75x + 125$	4 3 2 1	Graphs and/or equations all correct Minor error in changing Some comparison without creating equivalent representations One format is changed correctly.
Correct Solution: Easy Calculation Shirts: $20 \leq x \leq 40$ Buy More Shirts: $40 \leq x \leq 80$ Cool Shirts Inc.: $x \geq 80$	3	Award partial credit based on minor errors.
Clear explanation of the work and answer	3	
Part 2: A choice is made for how many T-shirts to order, a rationale for that choice is provided, and the correct selection based on that number is given.	5 3 1	All components are well written. Writing is marginal, rationale is not included, or incorrect choice is given A solution is provided, but it lacks any explanation.
Mathematics Practice: Select one of the listed practices to evaluate. This component could be evaluated from an observation of the student or team working.	3	Demonstration of the practice; Partial credit can be awarded.
Total Points	**18 points**	

Chapter 3 Performance Task (continued)

The Cost of a T-Shirt

You receive bids for making T-shirts for your class fundraiser from four companies. To present the pricing information, one company uses a table, one company uses a written description, one company uses an equation, and one company uses a graph. How will you compare the different representations and make the final choice?

The design is ready, and your job is to select the best option for ordering the shirts. The proposals from the four companies are listed below. Another slight problem is that the class officers have not decided how many T-shirts you should order.

Part 1:

Compare your options and explain which company you would select for orders from 20 to 200 T-shirts. Explain your reasoning clearly by using the correct terminology. Remember, most of the other class officers have not yet taken this class, and your class sponsor is a mathematics teacher.

Buy More Shirts

Number of shirts	20	40	60	80	100
Price (dollars)	130	200	270	340	410

Number of shirts	120	140	160	180	200
Price (dollars)	480	550	620	690	760

Easy Calculation Shirts

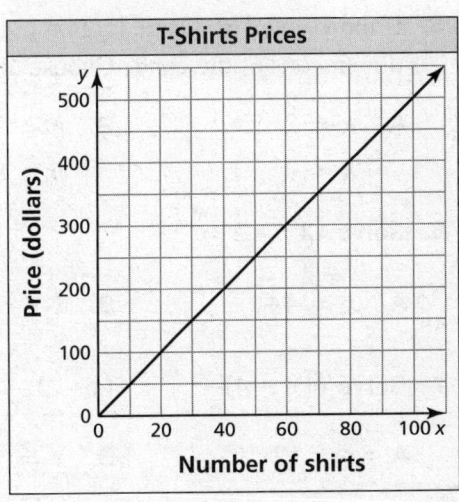

Cool Shirts, Inc.

We have a very easy pricing structure. We require an initial fee of $125 to set up your screen print and get everything in place. Then the cost is only $2.75 per shirt.

T-Shirts R Us

$C(x) = 2x + 180$

Part 2:

Based on what you have discovered, create your proposal to the class officers. Include how many T-shirts you think you should order, the company you would select, and what your option would ultimately cost per T-shirt.

Chapters 1–3 Cumulative Test

1. Which equation does not belong with the other three?

   A. $x + 3 = 7$ B. $\dfrac{x}{2} = 2$ C. $x - 5 = 9$ D. $4x = 16$

2. Solve $-8r = 56$.

   A. $r = 64$ B. $r = -7$ C. $r = 7$ D. $r = -448$

3. Solve $\dfrac{3}{5}x = 15$.

   A. $x = 75$ B. $x = 9$ C. $x = 25$ D. $x = 45$

4. A basketball costs $15.00. If the sales tax is 6%, which equation can be used to find the amount of tax?

   A. $t = 0.06 \cdot 15$ B. $0.06t = 15$ C. $15 + 0.06 = t$ D. $t = \dfrac{15}{0.06}$

5. If the volume of a cone is 24π cubic inches, which of the following are possible dimensions of the cone? Choose all that apply.

   A. $r = 2$ B. $r = 3$ C. $r = 6$ D. $d = 4$
      $h = 4$ $h = 8$ $h = 2$ $h = 18$

6. Solve $12x + 2 = 26$.

   A. $x = 4\dfrac{1}{6}$ B. $x = 2\dfrac{1}{3}$ C. $x = 12$ D. $x = 2$

7. Solve $6(x - 4) + x = 5(x - 2)$.

   A. $x = 17$ B. $x = 7$ C. $x = -7$ D. no solution

Chapters 1–3 Cumulative Test (continued)

8. Translate the equation $4x - 5 = 3(x + 2)$ into a verbal sentence.

   A. Four times the difference of x and five equals three times the sum of x and two.

   B. Four times x less than five is the same as three times x plus two.

   C. The sum of four times x and five is three times the sum of x and two.

   D. The difference of four times x and five is three times the sum of x and two.

9. A boat travels upstream on the Allegheny River for 2 hours. The return trip only takes 1.7 hours because the boat travels 2.5 miles per hour faster downstream due to the current. How far does the boat travel one way?

   A. $14\frac{1}{6}$ mi

   B. 14 mi

   C. $28\frac{1}{3}$ mi

   D. 28 mi

10. Solve $\frac{1}{3}(6x + 30) = 4x + 2(x - 7)$.

    A. $x = 6$

    B. $x = 11$

    C. $x = 4\frac{1}{4}$

    D. $x = -1$

11. Solve $|3n + 6| = 21$.

    A. $n = -5$ and $n = 5$

    B. $n = -9$ and $n = 9$

    C. $n = -9$ and $n = 5$

    D. no solution

12. Translate $5x + 7 \geq 2$ into a verbal sentence.

    A. Five times the sum of x and seven is greater than two.

    B. Five times x, increased by seven, is less than or equal to two.

    C. The product of five and x, increased by seven, is greater than or equal to two.

    D. Seven more than the product of five and x is greater than two.

Chapters 1–3 Cumulative Test (continued)

13. Which inequality is different?

    A. x is no more than 12.

    B. x is less than or equal to 12.

    C. x is at most 12.

    D. x is no less than 12.

14. Translate "seven is greater than the difference of twice a number and nine" into an inequality. Solve the inequality.

    A. $7 > 2n - 9$; $n < 8$

    B. $7 \geq 2(n - 9)$; $n \leq 12.5$

    C. $2n - 9 > 7$; $n > 8$

    D. $2(n - 9) > 7$; $n > 12.5$

15. Solve $2t - 1 > 15$.

    A. $t > 7$

    B. $t < 7$

    C. $t > 8$

    D. $t < 8$

16. An error was made while solving the inequality $-4(2x - 3) > 5x - 14$. Choose the step where the error was made.

    A. $-8x + 12 > 5x - 14$

    B. $-13x + 12 > -14$

    C. $-13x > -26$

    D. $x > 2$

17. Which compound inequality is represented by the graph?

    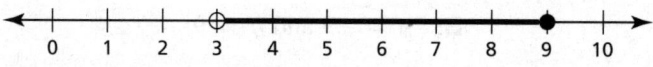

    A. $5 \leq x + 2 < 11$

    B. $10 < 2x + 4 \leq 22$

    C. $5 \leq x + 2 \leq 11$

    D. $10 \leq 2x + 4 < 22$

18. Solve $-3|2x - 5| \leq -9$.

    A. $x \geq 4$ or $x \leq 1$

    B. $x \leq 4$ and $x \geq 1$

    C. all real numbers

    D. no solution

19. Which of the following is a linear function?

    A. $y = x^2 + 4$

    B. $y = \sqrt{x} - 2$

    C. $y = \frac{1}{2}(x - 4)$

    D. $y = \frac{2}{x} + 3$

Chapters 1–3 Cumulative Test (continued)

20. If $f(x) = 3x + 12$, find the value of x so that $f(x) = 36$.

    A. 120   **B.** 16   **C.** 24   **D.** 8

21. The table shows the cost per hour to rent a canoe.

Number of hours, h	Rental fee, f
1	15
2	20
3	25
4	30

Which function represents the situation?

    A. $f(h) = 5h + 10$ **B.** $f(h) = 10h + 5$

    C. $f(h) = 5h + 15$ **D.** $f(h) = 10h + 15$

22. What is the equation of the graphed line, written in standard form?

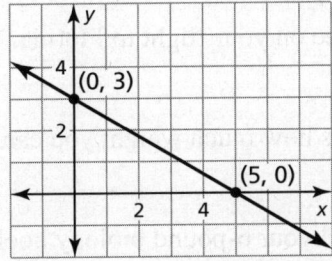

    A. $-3x - 5y = 15$ **B.** $y = -\frac{3}{5}x + 3$

    C. $3x + 5y = 15$ **D.** $y = -\frac{3}{5}(x - 5)$

23. A line contains the point $(0, 7)$ and has the same slope as the line $3x + y = 4$. Which of the following is the equation of the line in slope-intercept form?

    A. $3x + y = 7$ **B.** $y = -3x + 4$

    C. $y - 7 = -3x$ **D.** $y = -3x + 7$

Name _____ Date _____

Chapters 1–3 Cumulative Test (continued)

24. The function $f(x) = x$ is translated 4 units left and 3 units up. Which function represents the transformation?

   A. $g(x) = (x + 4) + 3$
   B. $g(x) = (x - 4) + 3$
   C. $g(x) = (x - 3) + 4$
   D. $g(x) = (x + 3) - 4$

25. Which equation does not belong with the other three? Explain your reasoning.

| $\lvert y \rvert = -8$ | $-5\lvert 2 + 3y \rvert = 15$ | $-3\lvert 1 - y \rvert = -9$ | $\lvert 2x + 4 \rvert + 8 = 2$ |

26. Which of the following equations are equivalent?

| $15 - 2x = 3x$ | $x + 2 = 5$ | $4(x - 2) = 3x - 5$ | $\frac{1}{3}x = 9$ |

27. The formula for the volume of a cylinder is $V = \pi r^2 h$.

   a. Solve the formula for h.

   b. Use the formula from part (a) to find the height of the cylinder with a radius of 3 centimeters and a volume of 45π cubic centimeters.

28. You are permitted a 50-pound check bag for free on your flight to Florida. Your bag weighs 32 pounds.

   a. Write and solve an inequality that represents how much weight you can add to your bag.

   b. Can you add your 5-pound algebra book and your 6-pound biology book without going over the weight limit? Explain.

29. Place each equation into one of the four categories.

No solution	One solution	Two solutions	Infinitely many solutions

| $8x + 12 = 4(2x + 3)$ | $\lvert x - 5 \rvert = 4$ | $\lvert 2x - 5 \rvert + 3 = 0$ | $\lvert 3x + 6 \rvert = 9$ |

| $3(-2 - 3x) = -9x - 4$ | $\frac{1}{2}(6 - x) = x$ | $1 - x = 6 - 6x$ | $3(x + 1) - 4 = 3x - 1$ |

Chapters 1–3 Cumulative Test (continued)

30. You want to average at least 90% on your algebra quizzes. So far, you have scored 93%, 97%, 81%, and 89% on your quizzes.

    a. What must you score on your next quiz to have at least a 90% average?

    b. Is it possible for you to raise your quiz average to 97% if only one more quiz is added? Explain your reasoning.

31. Certain amusement park rides have minimum and maximum height requirements. To ride the roller coaster in Kiddie Land, you must be between 36 and 54 inches tall.

    a. Write a compound inequality to represent the height requirements.

    b. If you are more than 54 inches tall, you must be accompanied by a child. Write an inequality to represent this requirement in feet.

32. The industry standard for ice cream storage is $-28.9°C$. Freezer temperatures fluctuate, so a safety factor of $2.8°C$ is allowed.

    a. Write and solve an absolute value inequality to find the minimum and maximum temperatures at which ice cream should be stored.

    b. What would be an advantage of setting the freezer temperature at the minimum temperature allowed? How could this possibly be a disadvantage?

33. Each mile you run, your body burns 100 calories.

    a. Write a function to represent this relationship.

    b. What is the domain and range of the function?

    c. Is the domain of the function discrete or continuous? Explain your reasoning.

34. Which equations are equivalent to $y = mx + b$?

$y - b = mx$	$x = \dfrac{y - b}{m}$	$m = \dfrac{y - b}{x}$
$b = y - mx$	$\dfrac{y - m}{b} = x$	$\dfrac{y - x}{m} = b$

Name_____ Date_____

Chapter 4 Quiz
For use after Section 4.3

Write an equation of the line in slope-intercept form.

1.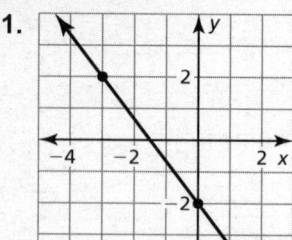
2.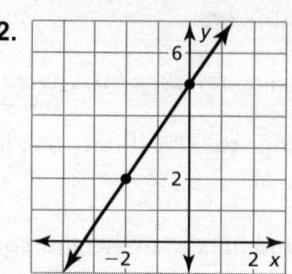
3.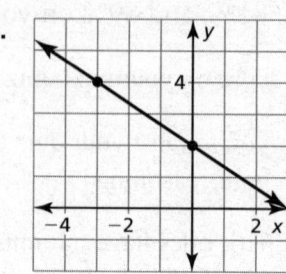

Write an equation in point-slope form of the line that passes through the given points.

4. $(4, 1), (-2, 7)$
5. $(1, 3), (-3, 0)$
6. $(-2, -5), (4, -1)$

Write a linear function f with the given values.

7. $f(0) = 2, f(3) = -1$
8. $f(-4) = -5, f(2) = -3$

Determine which of the lines, if any, are parallel or perpendicular. Explain.

9. The lines pass through the following points:
   Line a: $(1, 5)$ and $(-2, -4)$
   Line b: $(3, 2)$ and $(1, -4)$
   Line c: $(6, 1)$ and $(-4, 2)$

10. Line a: $4y - 12x = 20$
    Line b: $3y = -x + 2$
    Line c: $y = 3x - 1$

11. To rent office space for your business, you must pay a one-time fee of $1000 and pay rent of $800 per month.

    a. Write a linear model that represents the total cost of renting office space as a function of the number of months you will rent.

    b. Find the total cost of renting office space for one year.

    c. A different building has office space for rent that does not require a one-time fee, but you must pay rent of $1000 per month. If you have $15,000, at which building can you rent office space for the greatest amount of time? Explain.

12. The table shows the distance covered by a spaceship in outer space. Can the situation be modeled by a linear equation? Explain. If possible, write a linear model that represents the distance traveled as a function of time.

Time (seconds)	1	4	7	10	13
Distance (miles)	5	20	35	50	65

Answers

1. _____
2. _____
3. _____
4. _____
5. _____
6. _____
7. _____
8. _____
9. _____
   _____
   _____
10. _____
    _____
    _____
11. a._____
    b._____
    c._____
    _____
    _____
12. _____
    _____
    _____

Name_____ Date_____

Chapter 4 Test A

State the domain and the range of the function.

1. $f(x) = \begin{cases} -\frac{1}{2}x + 2, & \text{if } x > 2 \\ -x - 2, & \text{if } x \leq 2 \end{cases}$

2. $g(x) = \begin{cases} -2, & \text{if } 1 \leq x < 3 \\ -x + 4, & \text{if } -1 \leq x < 1 \\ 3, & \text{if } x < -1 \end{cases}$

3. Graph the function.

   $h(x) = \begin{cases} -2x - 2, & \text{if } x > 0 \\ 1, & \text{if } x \leq 0 \end{cases}$

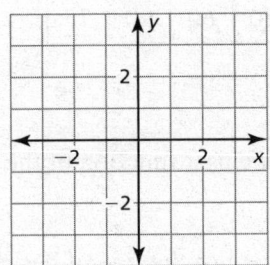

Answers

1. _____

   _____

2. _____

   _____

3. __See left.__

4. _____

5. _____

6. _____

7. _____

8. _____

9. _____

10. _____

11. _____

12. _____

Write the slope-intercept form of the equation with the given characteristics.

4. slope = $\frac{1}{4}$; y-intercept = 2

5. slope = $-\frac{3}{2}$; passes through $(-4, 7)$

6. passes through $(-2, 1)$ and $(2, -5)$

7. parallel to the line $y = -3x + 5$; passes through $(-4, 5)$

8. perpendicular to the line $y = \frac{1}{2}x - 8$; passes through $(7, -6)$

Write the point-slope form of the equation with the given characteristics.

9. slope = 2; y-intercept = 3

10. slope = -2; passes through $(-3, 5)$

11. parallel to the line $y = \frac{3}{5}x - 8$; passes through $(0, -3)$

12. perpendicular to the line $y = -2x - 7$; passes through $(-3, 10)$

Chapter 4 Test A (continued)

Determine if the sequence is arithmetic. If so, find the common difference.

13. 20, 13, 6, −1, …

14. 2, 4, 8, 16, …

15. −1, −5, −9, −13, …

16. 7, 4, 1, −1, …

17. The table shows the time x (in hours) students spent studying for a science exam and the grade they received.

Time (hours), x	3	2	5	1	0	4	3
Grade, y	84	77	92	70	60	90	75

   a. Describe the correlation.

   b. Write an equation that models grade as a function of the hours spent studying.

   c. Interpret the slope and the y-intercept of the line of best fit.

18. Consider the data used in Exercise 17.

   a. Use a graphing calculator to find an equation of the line of best fit.

   b. Identify and interpret the correlation coefficient.

   c. Predict the grade of a student who studied for 3 hours.

Determine if the given lines are parallel, perpendicular, or neither.

19. $y + 5 = -12$
    $x - y = 10$

20. $3x - 5y = 10$
    $10x + 6y = -36$

21. $2x - y = 10$
    $-4x - 2y = -8$

22. $-2x - 3y = 9$
    $4x + 6y = 24$

Tell whether a correlation is likely in the situation. Explain your reasoning.

23. the height of a person and the length of their stride

24. the number of flat tires on your car and the number of pets you own

25. the number of text messages sent daily and the number of meals eaten daily

Answers

13. _____
14. _____
15. _____
16. _____
17. a. _____
    b. _____
    c. _____
18. a. _____
    b. _____
    c. _____
19. _____
20. _____
21. _____
22. _____
23. _____
24. _____
25. _____

Name_____ Date_____

Chapter 4 Test B

State the domain and the range of the function.

1. $f(x) = \begin{cases} -\frac{3}{4}x - 1, & \text{if } x < 4 \\ 3, & \text{if } x \geq 5 \end{cases}$

2. $g(x) = \begin{cases} 4 - x, & \text{if } 1 < x < 4 \\ -2, & \text{if } -1 \leq x < 1 \\ 3, & \text{if } x < -1 \end{cases}$

3. Graph the function.

   $h(x) = \begin{cases} \frac{2}{3}x - 5, & \text{if } x > 0 \\ -\frac{1}{2}x - 3, & \text{if } x \leq 0 \end{cases}$

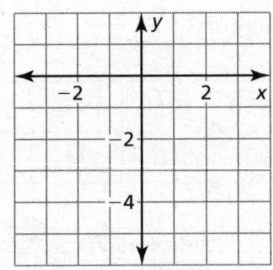

Write the slope-intercept form of the equation with the given characteristics.

4. slope = $\frac{2}{5}$; passes through $(-3, 1)$

5. passes through $(3, 5)$ and $(-1, 5)$

6. parallel to the line $2x - y = 7$; passes through $(-5, -3)$

7. perpendicular to the line $y = -\frac{3}{2}x - 7$; passes through $(-3, -4)$

8. perpendicular to the line $2x - 5 = -11$; passes through $(7, 5)$

Write the point-slope form of the equation with the given characteristics.

9. slope = $\frac{1}{2}$; x-intercept = 3

10. slope = -3; passes through $(4, -7)$

11. parallel to the line $2x - 5y = -20$; passes through $(7, 6)$

12. perpendicular to the line $y = 3x + 8$; passes through $(-4, 1)$

Determine if the sequence is arithmetic. If so, find the common difference.

13. $-3, -1, 3, 5, \ldots$

14. $-1, -7, -13, -19, \ldots$

15. $-\frac{1}{6}, \frac{1}{6}, \frac{1}{2}, \frac{5}{6}, \ldots$

16. $-1.2, -0.1, 0.8, 1.7, \ldots$

Answers

1. _____

2. _____

3. _See left._

4. _____

5. _____

6. _____

7. _____

8. _____

9. _____

10. _____

11. _____

12. _____

13. _____

14. _____

15. _____

16. _____

Chapter 4 Test B (continued)

17. Line m represents a translation of line ℓ 2 units up and 3 units right. Write an equation that represents the equation of line ℓ.

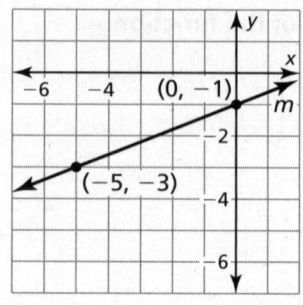

Answers

17. _____

18. a. __See left.__

b. _____

c. _____

18. The table shows the number of women (in millions) in the U.S. work force at various times during the past century.

Year, x	1900	1920	1930	1950	1970	1990
Number, y	5	8	10	16	31	57

a. Make a scatter plot of the data. Describe the correlation.

b. Use a graphing calculator to find an equation of the line of best fit.

c. Identify and interpret the correlation coefficient.

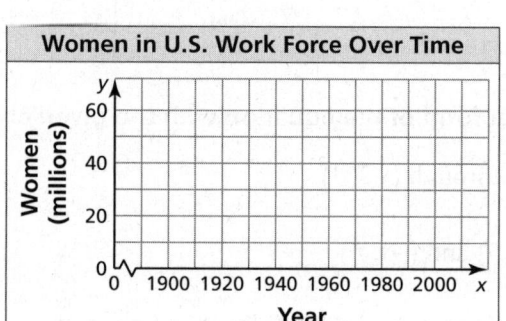

19. _____

20. _____

21. _____

22. _____

23. _____

Determine if the given lines are parallel, perpendicular, or neither.

19. $2x - 3y = 9$
$4x - 5y = 15$

20. $x = 5$
$2x - 3 = 15$

21. $2 - x = 3y$
$2y + 10 = 6x$

22. $y + x = \frac{1}{2}x + 1$
$2x - y = 3$

24. _____

Tell whether a correlation is likely in the situation. Explain your reasoning.

23. the amount of gas in a car's tank and the number of miles driven

24. the height of a person and the length of the person's hair

Name_____ Date_____

Chapter 4 Alternative Assessment

1. Use the given line to find the following.

   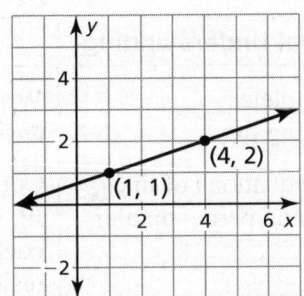

   a. Find the slope between the two points.

   b. Write an equation of the given line in point-slope form.

   c. Write an equation of the given line in slope-intercept form.

   d. Write an equation of the line that passes through $(3, -1)$ and is parallel to the given line.

   e. Write an equation of the line that passes through $(-2, 7)$ and is perpendicular to the given line.

2. The table shows the height x (in inches) and the number of strawberries y of several strawberry plants.

Height (inches), x	4	6	5	8	6	9	7	4	5	8
Strawberries, y	19	24	22	26	25	32	28	22	25	30

   a. Use a graphing calculator to find an equation of the line of best fit.

   b. Identify and interpret the correlation coefficient.

   c. Make a scatter plot of the residuals to verify that the model is a good fit.

3. $y = \begin{cases} -2x, & \text{if } x \leq 3 \\ \frac{1}{3}x - 2, & \text{if } x > 3 \end{cases}$

   a. Graph the function.

   b. Describe the domain and range.

   c. Evaluate the function when $x = 3$.

   d. Evaluate the function when $x = 0$.

   e. Evaluate the function when $x = 5$.

Name _____ Date _____

Chapter 4 Alternative Assessment Rubric

Score	Conceptual Understanding	Mathematical Skills	Work Habits
4	Shows complete understanding of: • finding equations of lines, parallel and perpendicular lines • line of best fit, correlation coefficient, residuals • piecewise functions	Writes equations of lines to answer all of Exercise 1 Correctly finds the line of best fit, interprets the correlation coefficient, and plots the residuals in Exercise 2 Correctly graphs, states the domain and range, and evaluates the piecewise function in Exercise 3	Answers all parts of all three problems All the equations and graphs are written or drawn carefully and systematically. Work is very neat and well organized.
3	Shows nearly complete understanding of: • finding equations of lines, parallel and perpendicular lines • line of best fit, correlation coefficient, residuals • piecewise functions	Writes equations of lines to answer most of Exercise 1 Correctly finds the line of best fit and the correlation coefficient, and plots the residuals in Exercise 2 Correctly graphs, states the domain and range, and evaluates some of the values of the piecewise function in Exercise 3	Answers several parts of the three problems Most of the equations and graphs are written or drawn carefully and systematically. Work is neat and organized.
2	Shows some understanding of: • finding equations of lines, parallel and perpendicular lines • line of best fit, correlation coefficient, residuals • piecewise functions	Writes equations of lines to answer some of Exercise 1 Correctly finds the line of best fit and the correlation coefficient in Exercise 2 Graphs some of the piecewise function, states the domain and range, and evaluates some of the values in Exercise 3	Answers some parts of the three problems The equations and graphs are written or drawn carelessly. Work is not very neat or organized.
1	Shows little understanding of: • finding equations of lines, parallel and perpendicular lines • line of best fit, correlation coefficient, residuals • piecewise functions	Writes some linear equations Attempts Exercise 2 Attempts Exercise 3 with few correct answers	Does not attempt many parts of the three problems Few equations or graphs are written or drawn. Work is sloppy and disorganized.

Name_____ Date_____

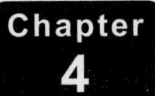 **Chapter 4 Performance Task**

Any Beginning

Instructional Overview	
Launch Question	With so many ways to represent a linear relationship, where do you start? Use what you know to move between equations, graphs, tables, and contexts.
Summary	There is a 3 by 3 matrix, and each cell asks the student to fill in different information about the same equation. Each problem will provide the information for a different cell, so the students will have a different starting place.
Teacher Notes	Model the problem, starting with $y = 2x - 5$. Have the students work in groups or guide them as they work through each cell. **Situation:** *Sample answer:* Your friend owes her mom $5, and she earns $2 for every day she makes her bed. **Table:** *Sample answer:* $(1, -3), (2, -1), (3, 1)$ **Slope-intercept form:** given **Graph:** **Standard form:** $-2x + y = -5$ **Points in functional notation:** *Sample answer:* $f(1) = -3$, $f(2) = -1$, $f(3) = 1$ **Point-slope form:** *Sample answer:* $y - 1 = 2(x + 3)$ **Equation of a parallel line:** *Sample answer:* $y = 2x + 1$ **Equation of a perpendicular line:** *Sample answer:* $y = -\frac{1}{2}x + 3$
Supplies	Copies of the matrix (4 per student), graph paper
Mathematical Discourse	If you are given the slope of a line, and a point on that line, where would you start to fill in the table? Where would you go next?
Writing/Discussion Prompts	1. Why do each of the different representations provide enough information to arrive at the other forms? 2. If you are given a choice on which cell to start with, which one would you select and why?

Name _____ Date _____

Chapter 4 Performance Task (continued)

Any Beginning

Curriculum Content	
Content Objectives	• Create equations in two variables to represent relationships between quantities and graph equations on coordinate axes. • Write equations in slope-intercept form and write an equation of a line given either its slope and a point on the line or two points on the line. • Write equations of parallel and perpendicular lines.
Mathematical Practices	• Recognize that a line is the same regardless of the method or structure used to represent it.

Rubric

Any Beginning	Points	
1: **Situation:** *Sample answer:* An international call requires $3 to connect and $0.25 for each additional minute. **Table:** given **Slope-intercept form:** $y = \frac{1}{4}x + 3$ **Graph:** 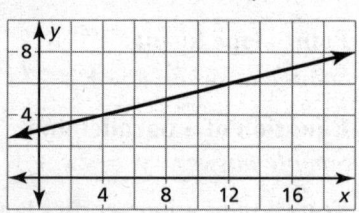 **Standard form:** $-x + 4y = 12$ **Points in functional notation:** *Sample answer:* $f(4) = 4$, $f(8) = 5, f(-4) = 2$ **Point-slope form:** *Sample answer:* $y - 4 = \frac{1}{4}(x - 4)$ **Equation of a parallel line:** *Sample answer:* $y = \frac{1}{4}x + 4$ **Equation of a perpendicular line:** *Sample answer:* $y = -4x - 1$	4 3 2 1	All cells correct Most cells correct, with minor errors in the remaining cells Some cells correct, but not all cells are attempted Only 1 or 2 new cells completed correctly

Chapter 4 Performance Task (continued)

Rubric (continued)

Any Beginning		Points	
2: **Situation:** *Sample answer:* The temperature is −1° Fahrenheit and dropping $\frac{2}{3}$° Fahrenheit each hour. **Table:** *Sample answer:* (0, −1), (3, −3), (6, −5) **Slope-intercept form:** $y = -\frac{2}{3}x - 1$ **Graph:** 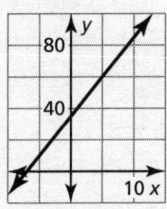	**Standard form:** $2x + 3y = -3$ **Points in functional notation:** *Sample answer:* $f(0) = -1,$ $f(3) = -3, f(6) = -5$ **Point-slope form:** *Sample answer:* $y + 3 = -\frac{2}{3}(x - 3)$	4	All cells correct
		3	Most cells correct, with minor errors in the remaining cells
		2	Some cells correct, but not all cells are attempted
		1	Only 1 or 2 new cells completed correctly
Equation of a parallel line: *Sample answer:* $y = -\frac{2}{3}x + 2$ **Equation of a perpendicular line:** *Sample answer:* $y = \frac{3}{2}x + 1$			
3: **Situation:** given **Table:** *Sample answer:* (1, 40), (2, 45), (3, 50) **Slope-intercept form:** $y = 5x + 35$ **Graph:**	**Standard form:** $-5x + y = 35$ **Points in functional notation:** *Sample answer:* $f(1) = 40,$ $f(2) = 45, \; f(3) = 50$ **Point-slope form:** *Sample answer:* $y - 40 = 5(x - 1)$ **Equation of a parallel line:** *Sample answer:* $y = 5x + 30$ **Equation of a perpendicular line:** *Sample answer:* $y = -\frac{1}{5}x + 10$	4	All cells correct
		3	Most cells correct, with minor errors in the remaining cells
		2	Some cells correct, but not all cells are attempted
		1	Only 1 or 2 new cells completed correctly

Name _____ Date _____

Chapter 4 Performance Task (continued)

Rubric (continued)

Any Beginning		Points	
4: **Situation:** Fencing for each rose bush is 4 feet and the contractor includes 3 feet for trimming errors. **Table:** *Sample answer:* $(1, 7), (2, 11), (3, 15)$ **Slope-intercept form:** $y = 4x + 3$ **Graph:** 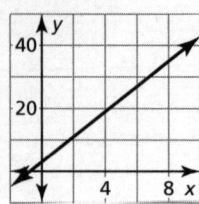 **Standard form:** given **Points in functional notation:** *Sample answer:* $f(1) = 7$, $f(2) = 11, f(3) = 15$ **Point-slope form:** *Sample answer:* $y - 7 = 4(x - 1)$ **Equation of a parallel line:** *Sample answer:* $y = 4x - 2$ **Equation of a perpendicular line:** *Sample answer:* $y = -\frac{1}{4}x + 2$		4 3 2 1	All cells correct Most cells correct, with minor errors in the remaining cells Some cells correct, but not all cells are attempted Only 1 or 2 new cells completed correctly
Mathematics Practice: When the students are able to move easily between the different representations and describe how they complete this process, they are embracing the linear structure.		4	The student demonstrates the ability to use the different representations, starting in any cell, and explains the path. Award partial credit as needed.
	Total Points	**20 Points**	

Performance Task (continued)

Any Beginning

With so many ways to represent a linear relationship, where do you start? Use what you know to move between equations, graphs, tables, and contexts.

Each cell of the matrix contains a different format for the same problem. Make four copies of the matrix. For each problem, start with the given information and fill in the rest of the chart. Each time, you will start in a different place.

1. Table: $\{(4, 4), (8, 5), (-4, 2)\}$

2. Slope and intercept: $m = \dfrac{-2}{3}$ and $b = -1$

3. Situation: The photography studio charges an initial fee of $35 and $5 for each picture.

4. Standard Form: $4x - y = -3$

Situation	Table x | y	Slope-intercept form
Graph	**Standard form**	**Points in functional notation**
Point-slope form	**Equation of a parallel line**	**Equation of a perpendicular line**

For each of the four problems, describe the order in which you filled in the matrix and why you chose that order.

Name _____ Date _____

Chapter 5 Quiz
For use after Section 5.4

Use the graph to solve the system of linear equations. Check your solution.

1. $y = x + 2$
   $y = -6x + 9$

2. $y = -\frac{1}{5}x + \frac{1}{5}$
   $y = 3x + 13$

3. $y = -\frac{1}{3}x - 2$
   $y = 4x - 2$

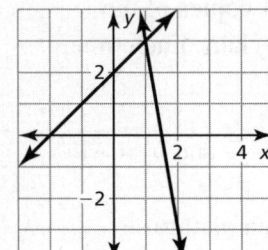
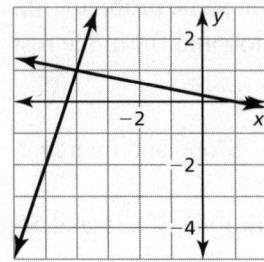
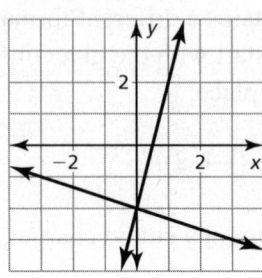

Solve the system of linear equations by substitution. Check your solution.

4. $y = 5x - 7$
   $-3x - 2y = -12$

5. $-4x + y = 6$
   $-5x - y = 21$

6. $-7x - 2y = -13$
   $x - 2y = 11$

Solve the system of linear equations by elimination. Check your solution.

7. $-4x - 2y = -12$
   $4x + 8y = -24$

8. $-2x - 9y = -25$
   $-4x - 9y = -23$

9. $-7x - 8y = 9$
   $-4x + 9y = -22$

Solve the system of linear equations.

10. $5x + 4y = 16$
    $-2x - 4y = -4$

11. $6x + 6y = -6$
    $5x + y = -13$

12. $-5x + y = -3$
    $3x - 8y = 24$

13. You spend $264 on clothes. Shirts cost $24 and pants cost $32. You buy a total of 9 items.

    a. Write a system of linear equations that represents this situation.

    b. Solve the system by graphing. Interpret your solution.

14. A phone company charges $0.06 per minute for local calls and $0.15 per minute for international calls. When your bill comes, it states that you accumulated 852 minutes with a charge of $69.84. Write and solve a system of linear equations to find the number of local and international minutes used.

Answers

1. _____
2. _____
3. _____
4. _____
5. _____
6. _____
7. _____
8. _____
9. _____
10. _____
11. _____
12. _____
13. a. _____

    b. _____

14. _____

Name_____ Date_____

Chapter 5 Test A

Solve the system of linear equations using any method.

1. $-6x + 5y = 1$
   $6x + 4y = -10$

2. $\frac{1}{2}x + y = -1$
   $y = \frac{1}{4}x - 4$

3. $-7x - 2y = -13$
   $x - 2y = 11$

4. $-5x + y = -3$
   $3x - 8y = 24$

5. $3x - 2y = 2$
   $5x - 5y = 10$

6. $6x + 6y = -6$
   $5x + y = -13$

Graph the inequality in a coordinate plane.

7. $x > -2$

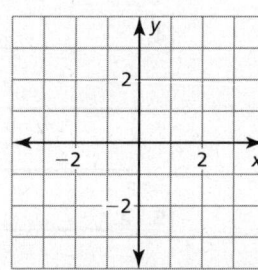

8. $y \leq -2x + 2$

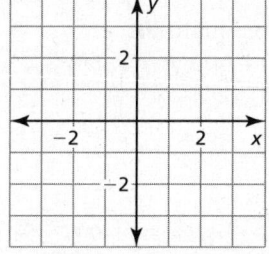

Answers

1. _____
2. _____
3. _____
4. _____
5. _____
6. _____
7. _See left._
8. _See left._
9. _See left._
10. _See left._
11. _____
_____
_____

Graph the system of linear inequalities.

9. $y < 3x - 4$
   $y \geq -\frac{1}{2}x + 3$

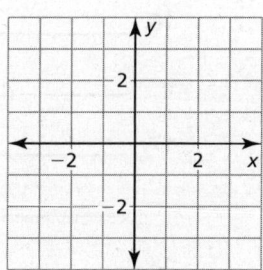

10. $3x - 2y \geq -2$
    $x - 2y < 2$

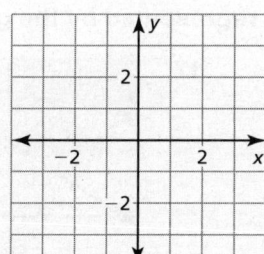

11. Two students are going to the store to buy school supplies for the new school year. One of the students buys 2 packs of pencils and 3 packs of pens for $8.25. Her friend purchases 5 packs of pencils and 2 packs of pens for $11.00. Is there enough information to determine the cost of 1 pack of pencils and 1 pack of pens? If so, find the cost of each.

Chapter 5 Test A (continued)

Compare the slopes and y-intercepts of the graphs of the equations in the linear system to determine whether the system has one solution, no solution, or infinitely many solutions. Explain.

12. $-3x + 3y = 4$
    $-x + y = 3$

13. $2x + 3y = -6$
    $-4x - 6y = 12$

14. $x + y = 7$
    $2x - 3y = -21$

15. You are buying plants and soil for your garden. The soil costs $4.00 per bag and the plants cost $10.00 each. You want to buy at least 5 plants and can spend no more than $100 total.

   a. Write a system of linear inequalities to model the situation.

   b. Graph the system of linear inequalities.

   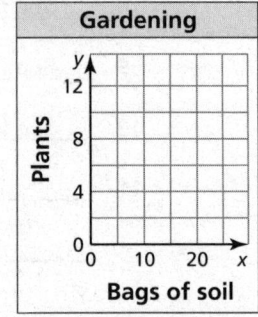

   c. Identify and interpret a solution to the system.

Write a system of linear inequalities represented by the graph.

16.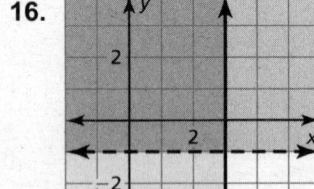

17.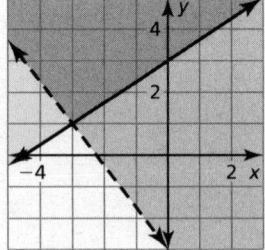

Solve the equation by graphing. Check your solutions.

18. $2x = -x + 3$

19. $2x - 1 = 5x + 5$

20. $|3x - 4| = |x|$

Answers

12. _____
    _____
    _____
13. _____
    _____
    _____
14. _____
    _____
    _____
15. a. _____
    _____
    b. __See left.__
    c. _____
    _____
16. _____
    _____
17. _____
    _____
18. _____
19. _____
20. _____

Name_____ Date _____

Chapter 5 Test B

Solve the system of linear equations using any method.

1. $x - 5y = -30$
   $3x + 5y = 10$

2. $x + 2y = -3$
   $-5x + 2y = 51$

3. $-5x - 4y = -15$
   $10x + 8y = 30$

4. $y = 2x + 3$
   $-4x + 2y = 8$

5. $y = -5x + 6$
   $2x + y = 6$

6. $x = -y - 1$
   $-5x + 2y = -65$

Graph the inequality in a coordinate plane.

7. $y > 0$

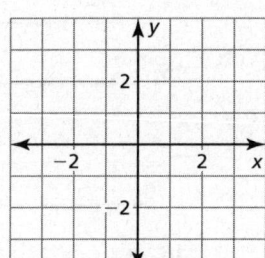

8. $2x - 5y \leq -10$

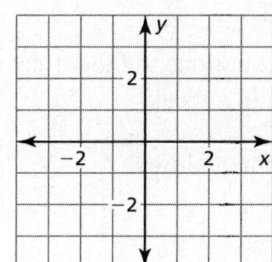

Graph the system of linear inequalities.

9. $3x + 2y \geq -2$
   $x + 2y \leq 2$

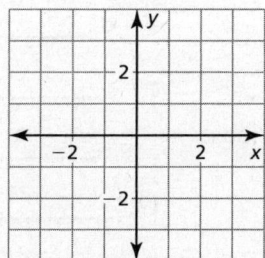

10. $2x - 3y \geq 6$
    $-4x + 6y \leq -18$

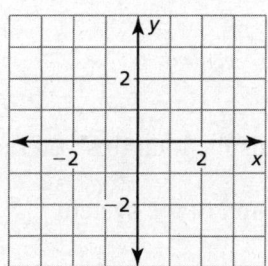

Answers

1. _____
2. _____
3. _____
4. _____
5. _____
6. _____
7. _See left._
8. _See left._
9. _See left._
10. _See left._
11. _____
12. _____

11. Write an expression that you can substitute for x in the top equation of the system below to solve the system by substitution.

    $5x - 2y = 8$
    $x - y = 1$

12. You have $8.80 in pennies and nickels. You have twice as many nickels as pennies. Write a system of linear equations that models the situation. How many of each type of coin do you have?

Name _____ Date _____

Chapter 5 Test B (continued)

Compare the slopes and *y*-intercepts of the graphs of the equations in the linear system to determine whether the system has one solution, no solution, or infinitely many solutions. Explain.

13. $x = -3y + 28$
    $x + 4y = 36$

14. $2x + 3y = 11$
    $-4x - 6y = -22$

15. $x + 2y = 3$
    $-2x - 4y = -20$

16. You make $5 an hour in tips working at a video store and $7 an hour in tips working at a landscaping company. You must work at least 4 hours per week at the video store, and the total number of hours you work at both jobs in a week cannot be greater than 15.

    a. Write a system of linear inequalities to model the number of hours that you could work at each location in a week.

    b. Graph the system of linear inequalities.

    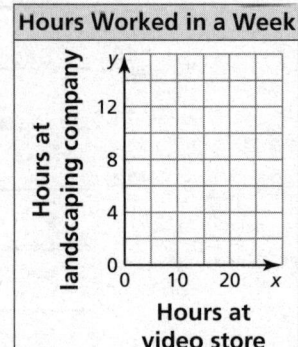

    c. Write an equation that models the total tips you receive from the two jobs.

    d. Identify and interpret a solution of the system.

Write a system of linear inequalities represented by the graph.

17.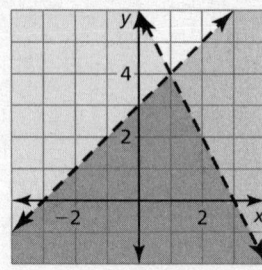

18.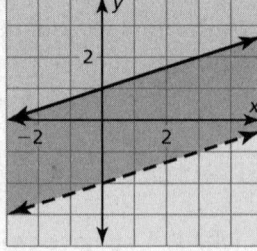

Solve the equation by graphing. Check your solutions.

19. $2x - 3 = x + 2$

20. $|x - 1| = |2x - 5|$

21. $|-x| = |2x - 3|$

Answers

13. _____
    _____
    _____

14. _____
    _____
    _____

15. _____
    _____
    _____

16. a. _____
    _____
    b. __See left.__
    c. _____
    d. _____
    _____

17. _____
    _____

18. _____
    _____

19. _____

20. _____

21. _____

68 Algebra 1
Assessment Book

Chapter 5 Alternative Assessment

1. Consider the system of linear equations

   $6x + 3y = -10$
   $-3x + y = 5$

   a. Solve the system by graphing.

   b. Solve the system by substitution.

   c. Solve the system by elimination.

   d. Which method do you prefer for this system? Explain.

   e. Change the coefficient of y in the first equation so that the system has infinitely many solutions.

2. Solve the equations by graphing. Check your solution.

   a. $-x + 5 = 4x - 5$

   b. $\frac{1}{4}x - 7 = -x + 3$

   c. $|x - 4| = |4x - 7|$

   d. $|3x - 1| = |-6x + 2|$

3. Consider the inequalities $y \leq 2x + 5$ and $x - y < -3$.

   a. Graph each inequality in a separate coordinate plane.

   b. Graph both inequalities in the same coordinate plane. Shade the intersection of the half-planes.

   c. Determine if $(0, 0)$ is a solution to the system of linear inequalities graphed in part (b).

   d. Determine if $(1, 4)$ is a solution to the system of linear inequalities graphed in part (b).

   e. Determine if $(1, 5)$ is a solution to the system of linear inequalities graphed in part (b).

Name _____ Date _____

Chapter 5 Alternative Assessment Rubric

Score	Conceptual Understanding	Mathematical Skills	Work Habits
4	Shows complete understanding of: • solving systems of linear equations • solving equations graphically by using a system of linear equations • graphing linear inequalities in two variables	Correctly solves a system of linear equations by graphing, substitution, and elimination Correctly solves equations graphically by using a system of linear equations Correctly graphs linear inequalities in two variables	Answers all parts of all three problems All equations and graphs are written or drawn carefully and systematically. Work is very neat and well organized.
3	Shows nearly complete understanding of: • solving systems of linear equations • solving equations graphically by using a system of linear equations • graphing linear inequalities in two variables	Solves a system of linear equations by two of the following methods: graphing, substitution, and elimination Correctly solves most of the equations graphically by using a system of linear equations Correctly graphs most of the linear inequalities in two variables	Answers several parts of the three problems Most equations and graphs are written or drawn carefully and systematically. Work is neat and organized.
2	Shows some understanding of: • solving systems of linear equations • solving equations graphically by using a system of linear equations • graphing linear inequalities in two variables	Solves a system of linear equations by one of the following methods: graphing, substitution, and elimination Correctly solves some of the equations graphically by using a system of linear equations Correctly graphs some of the linear inequalities in two variables	Answers some parts of the three problems The equations and graphs are written or drawn carelessly. Work is not very neat or organized.
1	Shows little understanding of: • solving systems of linear equations • solving equations graphically by using a system of linear equations • graphing linear inequalities in two variables	Does not solve a system of linear equations by one of the following methods: graphing, substitution, and elimination Correctly solves none of the equations graphically by using a system of linear equations Correctly graphs none of the linear inequalities in two variables	Does not attempt any part of any of the problems No equations or graphs are written or drawn. Work is sloppy and disorganized.

Name_____ Date_____

 **Performance Task**

Prize Patrol

Instructional Overview	
Launch Question	You have been selected to drive a prize patrol cart and place prizes on the competing teams' predetermined paths. You know the teams' routes and you can only make one pass. Where will you place the prizes so that each team will have a chance to find a prize on their route?
Summary	The students are given four equations they must graph and then find the equation of a line that must cross all four.
Teacher Notes	There are a number of options for the new equation, and rarely will the new equations have integer coordinates for the intersections. So, a graphing program on a tablet device or calculator might be helpful when completing this task. You can ask students to graph the first four equations before you allow them to use the graphing tool.
Supplies	Graphing calculator or other graphing tool, graph paper
Mathematical Discourse	What is going on in this game? What other rules might you place for the participants if you were running the game?
Writing/Discussion Prompts	1. If the first team that found its prize was the winner, would that change the game? 2. Would it matter which direction the team started on its path or where on the path it started?

Name _____ Date _____

Chapter 5 Performance Task (continued)

Prize Patrol

Curriculum Content	
Content Objectives	• Solve systems of linear equations by graphing. • Use systems of linear equations to solve real-life problems.
Mathematical Practices	• Use equations to model the paths that the participants would travel.

Rubric

Prize Patrol	Points	
1: [graph showing four lines labeled Team 1, Team 2, Team 3, Team 4 on coordinate plane]	8	Graphs all four equations correctly (2 points per correct graph)
2: The equation of the line is graphed correctly and intersects each of the four paths; *Sample answer:* $y = 7 - x$	4	The path created intersects all four equations. (1 point per intersection)
3: The four points of intersection with the new line are given and correct to the nearest tenth; *Sample answer:* Team 1: (1.5, 5.5), Team 2: (3.7, 3.3), Team 3: (6, 1), Team 4: (4.6, 2.4)	4	All ordered pairs are the correct point of intersection. (1 point per correct intersection)
4: There is a given explanation of the work and the choices made in the work. Some reasoning and an explanation for which team might find the prize first is included; *Sample answer:* It would depend which direction the teams travel, but if they are entering the park from the south, they should find them at approximately the same time.	3 2 1	Well written explanations Most explanation included with a few of the steps skipped Little explanation provided
Mathematics Practice: Students apply the equations of lines to describe real-world locations.	1	The student demonstrates an understanding that the equations of the lines are the paths the teams would travel. Award partial credit as needed.
Total Points	**20 points**	

72 Algebra 1
Assessment Book

Copyright © Big Ideas Learning, LLC
All rights reserved.

Name_____ Date_____

Performance Task (continued)

Prize Patrol

You have been selected to drive a prize patrol cart and place prizes on the competing teams' predetermined paths. You know the teams' routes and you can only make one pass. Where will you place the prizes so that each team will have a chance to find a prize on their route?

Here are the equations for each of the four competing teams' paths:

Team 1 Route: $x - 3y = -15$

Team 2 Route: $-5x + 2y = -12$

Team 3 Route: $x + 2y = -4$

Team 4 Route: $3x - 4y = 4$

Sketch the graphs of the four teams. Determine the equation for your prize patrol path and then identify the coordinates where you will place your prizes. Show your work and list the coordinates to the nearest tenth. Remember, all routes must be straight lines.

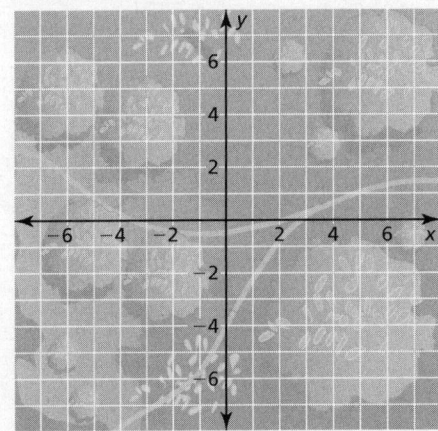

Prize patrol equation:	
Team	Location of the prizes
1	
2	
3	
4	

Explain how you decided on the location of your equation. Describe where another option might be. Based on where the prizes are hidden which team do you think might find the prize first and why? Explain a few things you had to consider when you made your choice.

Date _____

Cumulative Test

1. ...tion is represented by the graph?

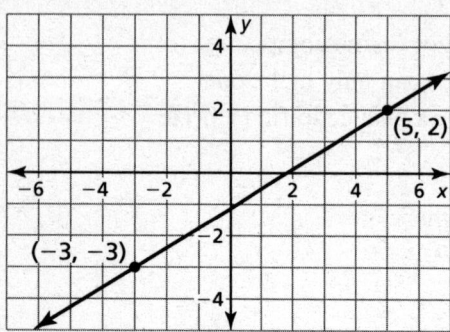

A. $y = \frac{5}{8}x - \frac{9}{8}$
B. $y = -\frac{8}{5}x + \frac{9}{5}$
C. $y = -\frac{5}{8}x + \frac{9}{8}$
D. $y = \frac{8}{5}x - \frac{9}{8}$

2. Use the numbers to fill in m and b in the equation $y = mx + b$ to represent the line in the graph.

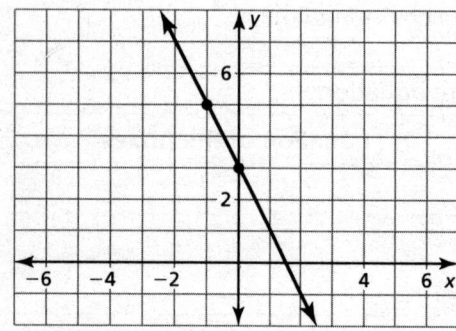

| −3 | −2 | 0 | 2 | 3 |

3. Which linear function f has the values $f(0) = 3$ and $f(-4) = 0$?

A. $f(x) = \frac{3}{4}x - 4$
B. $f(x) = \frac{3}{4}x + 3$
C. $f(x) = \frac{4}{3}x - 4$
D. $f(x) = \frac{4}{3}x + 3$

4. Which of the following is the equation of the line that passes through the point (4, 2) and has a slope of $\frac{1}{2}$? Select all that apply.

| $y - 4 = \frac{1}{2}(x - 2)$ | $y = \frac{1}{2}x$ | $y + 4 = \frac{1}{2}(x + 2)$ |
| $y - 2 = \frac{1}{2}(x - 4)$ | $x - 2y = 0$ | $y + 2 = \frac{1}{2}(x + 4)$ |

74 Algebra 1
Assessment Book

Name_____ Date _____

Chapters 4–5 Cumulative Test (continued)

5. Which of the following lines are perpendicular?

   | $y = -\tfrac{1}{3}x - 6$ | $y = -3x - 10$ | $y = 3x - 26$ | $y = -3x + 10$ |

6. Which formula represents the *n*th term of arithmetic sequence 2, 5, 8, 11, … ?

   A. $a_n = 3n + 5$ **B.** $a_n = 3n - 1$ **C.** $a_n = -7n + 5$ **D.** $a_n = -7n - 1$

7. Consider the function $f(x) = \begin{cases} 3x + 5, & \text{if } x \geq 4 \\ 2, & \text{if } 0 < x < 4 \\ x - 3, & \text{if } x \leq 0 \end{cases}$. Find $f(-2)$.

   A. −1   **B.** 1   **C.** 2   **D.** −5

8. If $y = -4x + 11$ and $3x + y = 9$, what is the value of *y*?

   A. 2   **B.** −2   **C.** 3   **D.** −1

9. The solution to which system of equations has a negative *x*-value?

   A. $y = x + 5$
      $3x + y = 25$

   B. $y = x + 2$
      $4x + y = 2$

   C. $x = y - 1$
      $-x + y = -1$

   D. $x = -1 - 2y$
      $3x + 4y = -3$

10. How many solutions does the system of linear equations $6a + 3b = 12$ and $6a - 3b = 12$ have?

    A. one   **B.** two   **C.** infinite   **D.** none

11. Jill owes her parents $100. She saves $25 every 2 weeks.

    a. Write the equation of the line that illustrates this situation.

    b. How long will it take her to pay her parents back the $100?

Chapters 4–5 Cumulative Test (continued)

12. Which inequality is represented by the graph?

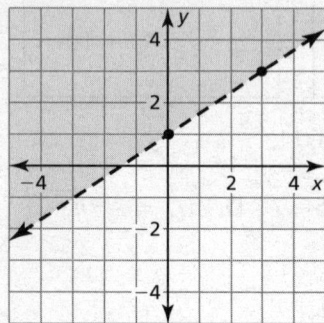

A. $y \geq \frac{2}{3}x + 1$ B. $y \leq \frac{2}{3}x + 1$ C. $y > \frac{2}{3}x + 1$ D. $y < \frac{2}{3}x + 1$

13. Which of the following is a solution of the system of inequalities shown below?

    $y > x - 3$
    $y \leq x + 3$

    A. $(0, 5)$ B. $(5, 4)$ C. $(1, -2)$ D. $(2, 6)$

14. Which of the following is *not* a solution of $2x - 3y \leq 12$?

    A. $(2, -5)$ B. $(0, 0)$ C. $(4, -1)$ D. $(-3, 4)$

15. Determine whether the sequence $x + 2, 2x + 2, 3x + 2, \ldots$ is arithmetic. If so, find the common difference and the next three terms. If not, explain your reasoning.

16. The graph represents a piecewise function.

    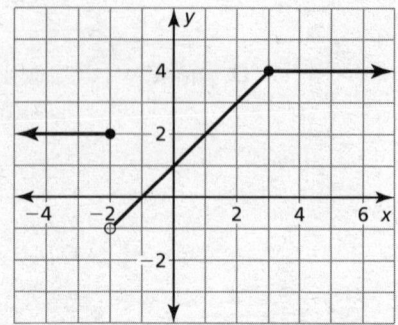

    a. Write a piecewise function for the graph.

    b. Use the graph to evaluate $f(0)$.

    c. Use the graph to evaluate $f(3)$.

Name_____ Date_____

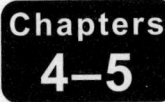
Cumulative Test (continued)

17. The following points are the vertices of a triangle: $(0, 4), (4, 0), (-4, 0)$.

    a. Write a system of linear inequalities so that the graph of the solutions of the system represents the triangle.

    b. Find the area of the triangle.

18. In 2014, the cost of a first class postage stamp was $0.49. The table shows the cost of mailing a letter of various weights.

Weight (ounces)	Cost (dollars)
1	0.49
2	0.70
3	0.91
3.5	1.12

    a. Write a function to represent the cost of mailing a stamped letter that weighs between 0 and 3.5 ounces.

    b. What type of function is this? Explain your reasoning.

    c. Graph the function.

19. A local pizza shop prices its pizza based on the number of toppings. The table shows the price p of a pizza with t toppings.

Number of toppings, t	Price, p
1	$11.25
2	$12.50
3	$13.75

    a. Write a function to represent the price p of a pizza in terms of the number t of toppings.

    b. What is the cost of a pizza with no toppings?

    c. A meat lover's pizza has bacon, sausage, pepperoni, and ham, and costs $14.50. How does this compare to a 4-topping pizza?

Copyright © Big Ideas Learning, LLC
All rights reserved.

Algebra 1
Assessment Book

Chapters 4–5 Cumulative Test (continued)

20. An algebra class collected the following data to determine if there is a correlation between height and shoe size.

Height (centimeters)	160	173	165	183	175	178
Shoe size	7	9	8.5	12	10	10

   a. Make a scatterplot using the data. Then draw a line of fit.

   b. Write an equation for the line of fit.

   c. Interpret the slope and the y-intercept.

   d. Determine whether the data show a *positive*, a *negative*, or *no* correlation.

   e. A basketball player is 7 feet 1 inch tall. Use your equation from part (b) to predict his shoe size.

   f. His actual shoe size is 22. How accurate is your equation?

   g. Use a graphing calculator to find the equation of the line of best fit.

   h. Identify and interpret the correlation coefficient.

21. You are in charge of the annual Powder Puff football game. You designed a T-shirt to sell during the event. A company charges $750 for the first 100 shirts and $500 for each additional 100 shirts.

   a. Write an equation that represents the total cost as a function of the number (in hundreds) of shirts ordered.

   b. Find the total cost of 500 shirts.

22. Which of the following lines are parallel to $3x + 2y = 6$? Select all that apply.

   | $y = -\frac{3}{2}x - 4$ | $3x - 2y = 6$ | $3x = 9 - 2y$ |

   | $y - 1 = -\frac{3}{2}(x + 2)$ | $y = \frac{3}{2}x + 5$ | $x = -\frac{2}{3}y + 2$ |

Chapters 4–5 Cumulative Test (continued)

23. You burn 20 calories per minute biking for x minutes and 10 calories per minute walking for y minutes. You spend a total of 90 minutes biking and walking and burn 1300 calories.

    a. Write a system of equations to determine how much time you spend on each exercise.

    b. How many minutes did you spend biking?

24. The Future Business Leaders of America club is setting up a school store. Members plan to sell pens and pencils. A pen is $1.25 and a pencil is $0.25. They would like to sell at least 50 pens and 100 pencils per week, with a goal of earning at least $115 per week.

    a. Define the variables and write a system of inequalities to represent the situation.

    b. Graph this system.

    c. Give two possible solutions to describe how the club can meet its goal.

    d. Is (55, 105) a solution? Explain.

25. Place each system of equations into one of the three categories.

No solution	One solution	Infinitely many solutions

$-8x - 10y = 24$
$6x + 5y = 2$

$-3x + 3y = 4$
$-x + y = 3$

$3 + 2x - y = 0$
$2y + 12 = 4x$

$5x + 4y = -14$
$3x + 6y = 6$

$7x + 2y = 18$
$6x + 6y = 0$

$2x + 8y = 6$
$-5x - 20y = -15$

Name _____ Date _____

Chapter 6 Quiz
For use after Section 6.4

Simplify the expression. Write your answer using only positive exponents.

1. $4^2 \cdot 4^3$
2. $(k^3)^{-2}$
3. $\left(\dfrac{4r^3}{2r^5}\right)$
4. $\left(\dfrac{5x^0}{10x^{-3}y^2}\right)^2$

Evaluate the expression.

5. $\sqrt[3]{64}$
6. $\left(\dfrac{1}{81}\right)^{1/4}$
7. $100^{3/2}$
8. $(\sqrt{9})^3$

9. Graph $y = 3^x$. Describe the domain and range.

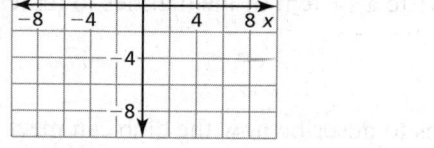

Answers

1. _____
2. _____
3. _____
4. _____
5. _____
6. _____
7. _____
8. _____
9. __See left.__
_____

Determine whether the table represents an *exponential growth function*, an *exponential decay function*, or *neither*. Explain.

10.
x	0	1	2	3
y	4	8	12	16

11.
x	1	2	3	4
y	28,561	2197	169	13

Determine whether the function represents *exponential growth* or *exponential decay*. Identify the percent rate of change.

12. $y = 4(1.65)^t$
13. $y = \dfrac{1}{4}(1.2)^t$
14. $f(t) = 70\left(\dfrac{2}{5}\right)^t$

15. The area of a rectangular yard with a width of $6a^2b$ feet is $72a^3b^4$ square feet. What is the length?

16. The function $f(t) = 16(1.4)^t$ represents the number of deer in a forest after t years.

   a. Does the function represent exponential growth or exponential decay?

   b. Graph the function on a calculator. Describe the domain and range.

   c. What is the yearly percent change?

   d. What is the approximate monthly percent change?

   e. How many deer are in the forest after 6 years?

10. _____
11. _____
_____
12. _____
_____
13. _____
_____
14. _____
_____
15. _____
16. a. _____
    b. _____
    c. _____
    d. _____
    e. _____

Name _____ Date _____

Chapter 6 Test A

Simplify the expression. Write your answer using only positive exponents.

1. $\dfrac{2^{-3} x^0}{y^{-4}}$

2. 3^0

3. $5^{-6} \cdot 5^3$

4. $(-4x^2)^3$

5. $\left(\dfrac{1}{3a^3}\right)^{-4}$

6. $12c^{-7}d^6$

Evaluate the expression.

7. $-\sqrt[3]{216}$

8. $(64)^{5/3}$

9. $(-8)^{2/3}$

Graph the function. Describe the domain and range.

10. $y = -2(3)^x$

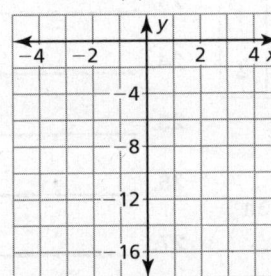

11. $y = 3(0.5)^x$

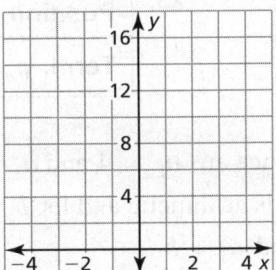

Solve the equation. Check your solution.

12. $3^x = \dfrac{1}{81}$

13. $25^{2x-3} = 125^{x+1}$

14. You deposit $500 in a savings account that earns 7% interest compounded annually.

   a. Write a function that represents the balance after t years.

   b. What is the balance after 2 years?

15. You buy a used car for $6599. Its value decreases by 12% every year.

   a. Write a function that represents the value y (in dollars) of the car after t years.

   b. What is the value of the car after 2.5 years?

   c. What is the value of the car after 20 years?

   d. According to the model, when will the value of the car be zero?

Answers

1. _____
2. _____
3. _____
4. _____
5. _____
6. _____
7. _____
8. _____
9. _____
10. ___See left.___
11. ___See left.___
12. _____
13. _____
14. a. _____
    b. _____
15. a. _____
    b. _____
    c. _____
    d. _____

Chapter 6 Test A (continued)

Determine whether the table represents an *exponential growth function*, an *exponential decay function*, or *neither*.

16.
x	1	2	3	4
y	2	8	24	128

17.
x	0	1	2	3
y	40	20	10	5

Decide whether the sequence is *arithmetic*, *geometric*, or *neither*.

18. 2, 4, 6, 8, …

19. 5, −10, 20, −40, …

20. 4, 9, 16, 25, …

21. −64, −32, −16, −8, …

Write a recursive rule for the sequence.

22.
Position, n	1	2	3
Term, a_n	25	10	−5

23.
Position, n	1	2	3	4
Term, a_n	−10	−6	−2	2

24. The first two terms of a sequence are $a_1 = 4$ and $a_2 = -2$. Let a_3 be the third term when the sequence is arithmetic and let b_3 be the third term when the sequence is geometric. Find $a_3 + b_3$.

Evaluate the function for the given value of x.

25. $y = 2^x$; $x = 5$

26. $f(x) = 3(4)^x$; $x = -1$

27. $f(x) = \frac{1}{2}(5)^x$; $x = 3$

28. $y = 0.5^x$; $x = -4$

29. The bacteria *E. coli* often cause illness among people who eat infected food. Suppose that a **single** *E. coli* bacterium in a batch of ground beef begins doubling every minute.

   a. Complete the table below that represents the number of bacteria after x minutes. (Assume no bacteria die.)

Minutes, x	0	1	2	3	4	5	6
Number of bacteria, y							

   b. Write an equation that can be used to calculate the number of bacteria in the food after any number of minutes.

   c. How many bacteria will there be after 20 minutes?

Answers

16. _____
17. _____
18. _____
19. _____
20. _____
21. _____
22. _____
23. _____
24. _____
25. _____
26. _____
27. _____
28. _____
29. a. __See left.__
    b. _____
    c. _____

Chapter 6 Test B

Simplify the expression. Write your answer using only positive exponents.

1. $\dfrac{12x^{-5}y^3}{3^{-2}x^{-2}y^{-4}}$
2. $(5x^4y^0)^{-3}$
3. $\left(-\dfrac{1}{2a^{-2}}\right)^{-3}$

Rewrite the expression as a power of a product.

4. $9x^6y^8$
5. $64x^9y^9$

Evaluate the expression.

6. $(27)^{-2/3}$
7. $-\sqrt[3]{-125}$
8. $(8)^{2/3} \cdot (27)^{-1/3}$

Graph the function. Describe the domain and range.

9. $y = -\dfrac{1}{2}(4)^x$

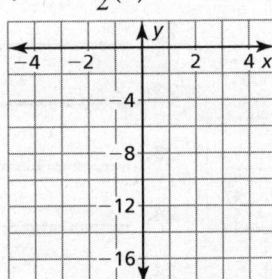

10. $y = -2(0.5)^x$

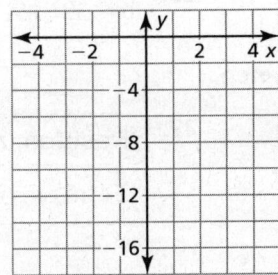

Answers

1. _____
2. _____
3. _____
4. _____
5. _____
6. _____
7. _____
8. _____
9. __See left.__
10. __See left.__
11. a. _____
    b. _____
12. a. _____
    b. _____
    c. _____

11. You deposit $675 in a savings account that earns 6% interest compounded monthly.

    a. Write a function that represents the balance after t years.

    b. What is the balance after 3 years?

12. You work for a design company and want to enlarge one of your designs for a new billboard sign. You can enlarge the design by 120% every time you run the program. The original design is 12 inches wide by 24 inches long.

    a. Write a function that represents the width after t times that you run the program.

    b. How wide is the design after running the program five times?

    c. How many times will you have to run the program for the design to be at least 150 feet wide?

Chapter 6 Test B (continued)

Solve the equation. Check your solution.

13. $2 \cdot 4^{x+1} = \frac{1}{32}$

14. $25^x \cdot 5^{x+3} = 625^{x-7}$

Determine whether the table represents an *exponential growth function*, an *exponential decay function*, or *neither*.

15.
x	1	2	3	4
y	2	−4	8	−16

16.
x	0	1	2	3
y	81	27	9	3

Determine whether the sequence is *arithmetic*, *geometric*, or *neither*.

17. 1, 3, 6, 10, …

18. −80, 40, −20, 10, …

19. $\frac{2}{3}, \frac{1}{6}, -\frac{1}{3}, -\frac{5}{6}, \ldots$

20. $\frac{1}{64}, \frac{1}{32}, \frac{1}{16}, \frac{1}{8}, \ldots$

Write a recursive rule for the sequence.

21.
Position, n	1	2	3
Term, a_n	27	9	3

22.
Position, n	1	2	3	4
Term, a_n	−2	−3	−5	−9

23. The first two terms of a sequence are $a_1 = 2$ and $a_2 = -1$. Let a_3 be the third term when the sequence is arithmetic and let b_3 be the third term when the sequence is geometric. Find $2a_3 - b_3$.

Evaluate the function for the given value of x.

24. $f(x) = \frac{1}{3}(4)^x; x = -2$

25. $f(x) = -2\left(\frac{1}{4}\right)^x; x = \frac{1}{2}$

26. The function $P(t) = 2000(1.5)^t$ represents the population of a small town.

   a. Does the function represent exponential growth or exponential decay?

   b. What is the yearly percent change in population?

   c. What is the approximate monthly percent change?

   d. How many people are living in the town after 3 years?

Answers

13. _____
14. _____
15. _____
16. _____
17. _____
18. _____
19. _____
20. _____
21. _____
22. _____
23. _____
24. _____
25. _____
26. a. _____
   b. _____
   c. _____
   d. _____

Chapter 6 Alternative Assessment

1. Simplify or evaluate the expression.

   a. $\dfrac{x^{-2}y^0}{x^4y^{-5}}$

   b. $\left(x^{-2} \cdot x^9\right)^{-3}$

   c. $\left(\dfrac{7x^5}{4y^3}\right)^2$

   d. $\sqrt[4]{256}$

   e. $(-27)^{2/3}$

2. Solve the equation.

   a. $\dfrac{1}{8} = 2^{2x-1}$

   b. $9^{4x-3} = 27^{x+3}$

   c. $\left(\dfrac{1}{4}\right)^{5x} = 32^{2x+8}$

3. State whether the sequence is defined *explicitly* or *recursively*. Then determine whether the sequence is *arithmetic*, *geometric*, or *neither*. Write the first six terms of the sequence. Then graph the sequence.

   a. $a_n = 10(2)^{n-1}$

   b. $a_n = -6 + (n-1)5$

   c. $a_n = 2\left(\dfrac{1}{3}\right)^{n-1}$

   d. $a_1 = 3,\ a_n = a_{n-1} + 2$

   e. $a_1 = 24,\ a_n = -\dfrac{1}{2}a_{n-1}$

Name_____ Date _____

Chapter 6 Alternative Assessment Rubric

Score	Conceptual Understanding	Mathematical Skills	Work Habits
4	Completely understands: • exponential and radical expressions • solving exponential equations • geometric and recursive sequences	Correctly simplifies or evaluates all of the exponential and radical expressions Correctly rewrites each side to a common base and solves for x for all of Exercise 2 Identifies all arithmetic, geometric, and recursive sequences, and writes all six terms	Answers all parts of the three problems All expressions and graphs are written or drawn carefully and systematically. Work is very neat and well organized.
3	Shows nearly complete understanding of: • exponential and radical expressions • solving exponential equations • geometric and recursive sequences	Correctly simplifies or evaluates most of the exponential and radical expressions Correctly rewrites each side to a common base and solves for x for most of Exercise 2 Identifies most of the arithmetic, geometric, and recursive sequences, and writes most of the terms	Answers several parts of both problems Most expressions and graphs are written or drawn carefully and systematically. Work is neat and organized.
2	Shows some understanding of: • exponential and radical expressions • solving exponential equations • geometric and recursive sequences	Correctly simplifies or evaluates some of the exponential and radical expressions Correctly rewrites each side to a common base and solves for x for some of Exercise 2 Identifies some of the arithmetic, geometric, and recursive sequences, and writes some of the terms	Answers some parts of both problems Expressions and graphs are written or drawn carelessly. Work is not very neat or organized.
1	Shows little understanding of: • exponential and radical expressions • solving exponential equations • geometric and recursive sequences	Does not correctly simplify or evaluate the exponential and radical expressions Does not rewrite each side to a common base and solves for x for Exercise 2 Identifies none of the arithmetic, geometric, and recursive sequences, and writes none of the terms	Does not attempt any part of either problem No equations or graphs are written or drawn. Work is sloppy and disorganized.

Name_____ Date_____

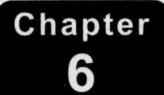 **Performance Task**

The New Car

Instructional Overview	
Launch Question	There is so much more to buying a new car than the purchase price. Interest rates, depreciation, and inflation are all factors. So, what is the real cost of your new car?
Summary	Buying a car is a stressful decision. There are so many different factors to consider. This assessment uses multiple factors in car purchases to help students interpret the real costs.
Teacher Notes	Students should use the Internet to find a car that interests them. There are choices for loans. Students could go to a local bank or credit union to investigate interest rates or you could provide the rate and number of years you want them to use.
Supplies	Handouts, calculators, Internet
Mathematical Discourse	Do you think it is a good idea to buy a new car? Why or why not?
Writing/Discussion Prompts	1. What are the extra costs you will need to expect when buying a car? 2. Do you think that buying a new car with a loan is a good idea? Why or why not?

Name _____ Date _____

 **Performance Task** (continued)

The New Car

Curriculum Content	
Content Objectives	• Use and interpret exponential growth and decay functions. • Construct a geometric sequence.
Mathematical Practices	• Investigate some hidden costs that occur when purchasing a car.

Rubric

The New Car		Points
1 Students select a car and three different option costs.	5 1	Total possible points car selection, each option up to and delivery costs
2–4 Sales tax calculated correctly, formula used to estimate total cost used correctly	4 1 2 1	Total possible points point for sales tax points monthly cost estimate total cost estimate
5 Depreciation Chart calculations with additional amount filled in for each year that matches with the percent value of the car	5 3 1	All answers correct Half the answers correct One or two of the calculations correct
6 The new price with inflation is calculated.	2	Cost of the car with inflation correct
7 The student makes some relevant comments about the purchase of a car that are supported by one or more previous calculations.	4 2	Thoughtful response with calculations having input on the response Response is well written without reference to the calculations
Mathematics Practice: Model with mathematics. Purchasing a car has many more factors than students ever consider.	3	It is clear through the student responses that they recognize the importance mathematics plays in the calculations.
Total Points	23 points	

Name_____ Date_____

Performance Task (continued)

The New Car

There is so much more to buying a new car than the purchase price. Interest rates, depreciation, and inflation are all factors. So, what is the real cost of your new car?

The Car

1. Use the Internet to look up the base purchase price for the car you would like to buy. Review the optional equipment and decide on three things you would like to add to your car. Add the destination and handling charges if you can find them.

Car Brand/options	Cost of the item
Destination and Handling	
Total Car Cost	

2. Next find the sales tax rate for your area. The rate is _____. Use that tax rate to calculate the cost of your car including sales tax.

3. Now, if you do not have that much money, you will need to borrow the money. There are many options available, with charts and a complex formula to calculate the exact payment, but you can estimate your payment more easily as follows.

   P = Principle (amount borrowed) = _____

   Y = Determine how many years you will take to pay (3 and 7 years). = _____

   R = interest rate you could borrow your money (2.5% and 8%) as a decimal
   = _____

   Estimate your monthly payment with the following formula.

   $$\frac{[(Y \cdot R) + 1] \cdot P}{12 \cdot Y} = \text{Monthly Payment} = \underline{\qquad}$$

Name _____ Date _____

 **Performance Task** (continued)

The New Car

4. The numerator of the formula in Exercise 3 is $[(Y \cdot R) + 1] \cdot P$. This value is an estimate for the amount you will ultimately pay back. Write that amount here
_____.

5. There are so many other costs people do not always prepare for or expect. Create a depreciation chart based on the amount you would have paid for your car from Exercise 1.

Estimated percent of value after depreciation	Amount of depreciation from previous time	Value of the car
New (100%)		(100%)
After 6 months		(90%)
After 1 year		(81%)
After 2 years		(65.61%)
After 3 years		(53.14%)
After 4 years		(43.05%)
After 5 years		(34.87%)

6. The inflation rate in February 2014 is approximately 1.6%. Traditionally, the U.S. averages 3% inflation. If you were to postpone buying the car for 5 years, what would be the estimated cost of the car? Use the formula:

   Price $(1 + 0.03)^5$ = new cost.

7. Considering all this information, what do you think would be your wisest choice for purchasing a car? Use the information and calculations from this activity to support your decision.

Performance Task (continued)

Teacher Notes:

Name _____ Date _____

Chapter 7 Quiz
For use after Section 7.4

Write the polynomial in standard form. Identify the degree and leading coefficient of the polynomial. Then classify the polynomial by the number of terms.

1. $-6r^4$

2. $4 + g^2 - 2g$

3. $\frac{1}{4}n^3 - \frac{3}{8}n^5$

4. $-1.6a + 2a^4 + 8.1a^3$

Find the sum or difference.

5. $(3x^2 + 7) + (-x^2 + 2)$

6. $(-4n^2 + 2n) - (n^2 - 5)$

7. $(-2h^2 + 2h) - (2h^2 - 5h + 12)$

8. $(m^2 - 2mn + n^2) + (-2m^2 + mn)$

Find the product.

9. $(x + 5)(x + 4)$

10. $(4 - 2d)(3d - 7)$

11. $(y + 6)(y^2 + 3y - 4)$

12. $(5z - 3)(5z + 3)$

Solve the equation.

13. $3x^2 - 12x = 0$

14. $(6 - y)(6 - y) = 0$

15. $(4p + 3)(2p - 5)(p + 2) = 0$

16. $-5y(y - 9)(3y + 2) = 0$

17. You are framing a picture with a frame of equal width on each side.

   a. Write a polynomial that represents the perimeter of the picture including the frame.

   b. Write a polynomial that represents the area of the picture including the frame.

   c. Find the perimeter and the area of the picture including the frame when the width of the frame is 2 inches.

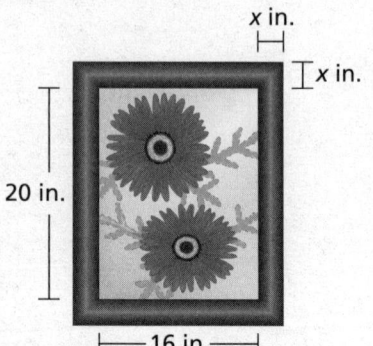

Answers

1. _____
2. _____
3. _____
4. _____
5. _____
6. _____
7. _____
8. _____
9. _____
10. _____
11. _____
12. _____
13. _____
14. _____
15. _____
16. _____
17. a. _____
    b. _____
    c. _____

Name_____ Date_____

Chapter 7 Test A

Write the polynomial in standard form. Identify the degree and leading coefficient of the polynomial. Then classify the polynomial by the number of terms.

1. $5 + x^2 - 7x$
2. $-7v^4$
3. $5z^6 - 1.5z^7 + z$
4. $-\frac{1}{2}a^4 + \frac{2}{3}a^5$

Find the sum or difference.

5. $(2b^3 - 5b) - (7b + 3b^2)$

6. $(-5x^2 - 2) + (3x^2 + 7)$

7. $(-x + 3) + (-11x^2 - 8 + 12x)$

8. $(y^2 + 4 - 2y^3) - (3y^2 - 7y + 4y^3)$

Find the product.

9. $(a + 5)(a - 3)$
10. $(2c - 1)(4 + 3c)$
11. $(7 + x)(x^2 + 6x - 8)$
12. $(4p - 3)^2$

13. You are making a fence for your garden. The length is five less than two times the width.

    a. Write a polynomial that represents the perimeter of the garden.

    b. Write a polynomial that represents the area of the garden.

    c. Find the perimeter and area of the garden when the width is 8 feet.

Factor the polynomial completely.

14. $a^2 + 10a + 16$
15. $5y^2 - 20$
16. $4s^2 - 1$
17. $3b^3 - 13b^2 + 10b$
18. $10x^3 - 12x^2 + 5x - 6$
19. $16t^2 - 24t + 9$

Answers

1. _____
   _____
2. _____
   _____
3. _____
   _____
4. _____
   _____
5. _____
6. _____
7. _____
8. _____
9. _____
10. _____
11. _____
12. _____
13. a. _____
    b. _____
    c. _____
14. _____
15. _____
16. _____
17. _____
18. _____
19. _____

Name _____ Date _____

Chapter 7 Test A (continued)

Find the x-coordinates of the points where the graph crosses the x-axis.

20.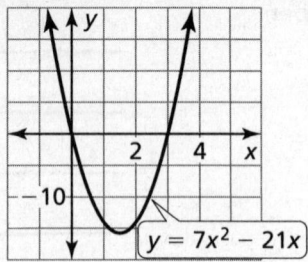

21.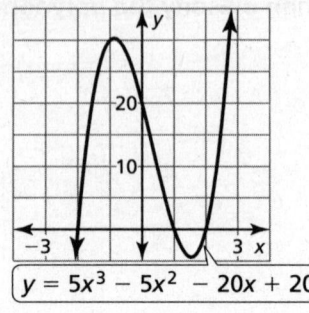

Solve the equation.

22. $(x - 2)(3x - 1) = 0$

23. $20m^2 - 40m = 0$

24. $4r^2 + 13r + 3 = 0$

25. $b^2 - 11b = -10$

26. The side of a cube is represented by x feet.

    a. Write a polynomial that represents the volume of the cube.

    b. Write a polynomial that represents the surface area of the cube.

    c. You increase each side of the cube by 1 foot. Write a polynomial that represents the new volume of the cube.

    d. You double the length of each side of the cube. Write a polynomial that represents the new surface area of the cube.

27. Write a polynomial that has two negative roots and one positive root.

28. The length of a rectangle in terms of x is $3x^2 + 7x + 9$ and its width is $3x + 6$.

    a. Write a polynomial that represents the perimeter of the rectangle.

    b. What is the perimeter when $x = 5$ feet?

Answers

20. _____
21. _____
22. _____
23. _____
24. _____
25. _____
26. a. _____
    b. _____
    c. _____
    d. _____
27. _____
28. a. _____
    b. _____

Name _____ Date _____

Chapter 7 Test B

Write the polynomial in standard form. Identify the degree and leading coefficient of the polynomial. Then classify the polynomial by the number of terms.

1. $3x + 5x^2 - 7x + 5$
2. $-5b^3$
3. $20z^4 - z^7 + \frac{2}{3}z$
4. $-\frac{3}{7}a^2 + \frac{4}{5}a^5$

Find the sum or difference.

5. $-5(4b^3 - 5b) - (3b^4 - b^3)$
6. $(-3x + 4x^2) + (-12x^3 + 7x)$
7. $(2 - 3x) + (14x - 7x^2 - 5)$
8. $(5 - 6y - 4y^2) - (-2y^2 + 5y + 12y^3)$

Find the product.

9. $(5 - a)(a^2 - 3a - 10)$
10. $(1 - 5c)(2c + 6)$
11. $2x(3x + 1)(x^2 + 4x)$
12. $5(p - 3)^2$

13. A rectangular picture is 6 centimeters longer than it is wide. A frame 1 centimeter wide is placed around the picture.

   a. Write a polynomial that represents the perimeter of the frame.
   b. Write a polynomial that represents the area of the frame.
   c. Find the perimeter of the frame if the picture is 15 centimeters wide?

Factor the polynomial completely.

14. $10nm^3 - 15n^2m$
15. $4x^2 + 2xy - 2y^2$
16. $3p^3 + 9p^2 - 210p$
17. $-5x^2 + 15x + 140$
18. $5x^3 - 125x$
19. $6ab + 12a^2 - 7xb - 14xa$

Answers

1. _____
   _____
2. _____
   _____
3. _____
   _____
4. _____
   _____
5. _____
6. _____
7. _____
8. _____
9. _____
10. _____
11. _____
12. _____
13. a. _____
    b. _____
    c. _____
14. _____
15. _____
16. _____
17. _____
18. _____
19. _____

Name _____ Date _____

Chapter 7 Test B (continued)

Find the x-coordinates of the points where the graph crosses the x-axis.

20.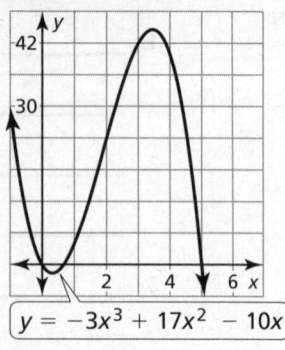

21.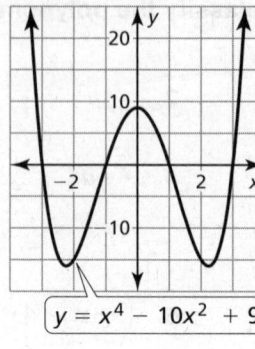

Answers

20. _____

21. _____

22. _____

23. _____

24. _____

25. _____

26. _____

27. a. _____

b. _____

Solve the equation.

22. $2p^2 + 24 = -16p$

23. $9m^2 - 1 = 0$

24. $-3x^2 + x^3 = 4x$

25. $x^4 - 5x^2 + 4 = 0$

26. An object is launched at 9.8 meters per second from a 73.5-meter tall platform. The object's height s (in meters) after t seconds is given by the equation $s(t) = -4.9t^2 - 9.8t + 73.5$. When does the object strike the ground?

27. You are designing an aquarium for your new office. The dimensions of the aquarium are restricted as shown in the diagram below.

   a. Write a polynomial expression that represents the volume of the aquarium according to the specified dimensions.

   b. You need the aquarium to hold 17,640 cubic inches of water. Find the possible dimensions of the aquarium.

Chapter 7 Alternative Assessment

1. Factor the polynomial completely.

   a. $q^2 + q - 30$

   b. $5x^2 - 17x - 12$

   c. $-2n^2 + 15n - 7$

   d. $a^2 - 121$

   e. $9k^2 + 24k + 16$

   f. $4x^3 - 12x^2 + 3x - 9$

   g. $p^3 - 49p$

2. Solve the equation.

   a. $(k - 8)(k + 2)(k - 7) = 0$

   b. $m^2 + 81 = 18m$

   c. $3y^4 + 27y^3 + 60y^2 = 0$

   d. $(x + 10)(x^2 - 25) = 0$

   e. $p^3 - 6p^2 - 4p + 24 = 0$

3. You are building a brick patio. The length of the patio is $(x + 11)$ feet.

   a. The total area of the patio can be represented by $x^2 + 14x + 33$. Write an expression for the width of the patio.

   b. Write an expression for the perimeter of the patio.

   c. The area of the patio is 240 square feet. Find the value of x.

   d. You are putting edging around the patio and have a budget of $200 for the edging. You found edging that costs $2.75 per foot. Is this edging within your budget? Explain.

Name _____ Date _____

Chapter 7 Alternative Assessment Rubric

Score	Conceptual Understanding	Mathematical Skills	Work Habits
4	Completely understands: • factoring polynomials • solving polynomial equations in factored form • using polynomials to solve perimeter and area problems	Completely factors all of the polynomials in Exercise 1 Correctly solves all of the polynomial equations in factored form Correctly uses polynomials to solve all of the perimeter and area problems	Answers all parts of all three problems All equations and polynomials are written carefully and systematically. Work is very neat and well organized.
3	Shows nearly complete understanding of: • factoring polynomials • solving polynomial equations in factored form • using polynomials to solve perimeter and area problems	Completely factors most of the polynomials in Exercise 1 Correctly solves most of the polynomial equations in factored form Correctly uses polynomials to solve most of the perimeter and area problems	Answers several parts of all three problems Most equations and polynomials are written carefully and systematically. Work is neat and organized.
2	Shows some understanding of: • factoring polynomials • solving polynomial equations in factored form • using polynomials to solve perimeter and area problems	Completely factors some of the polynomials in Exercise 1 Correctly solves some of the polynomial equations in factored form Correctly uses polynomials to solve some of the perimeter and area problems	Answers some parts of all three problems Equations and polynomials are written carelessly. Work is not very neat or organized.
1	Shows little understanding of: • factoring polynomials • solving polynomial equations in factored form • using polynomials to solve perimeter and area problems	Factors one of the polynomials in Exercise 1 Solves one of the polynomial equations in factored form Uses polynomials to solve perimeter and area problems	Does not attempt many parts of the three problems Equations and polynomials are written carelessly. Work is sloppy and disorganized.

Name_____ Date _____

 **Performance Task**

The View Matters

Instructional Overview	
Launch Question	The way an equation or expression is written can help you interpret and solve problems. Which representation would you rather have when trying to solve for specific information? Why?
Summary	Students are to look at different ways to represent the information provided in this chapter and select which representation might be best to help them find the solution.
Teacher Notes	Before the activity, encourage students to look through the modeling problems in their textbook to help remind them of the formats for the applications they have competed. In some cases, the students might have a good reason why a different representation was selected than the one provided in the grading rubric. Make sure that the students spend some time working with peers on problem 7 so that they have a chance to defend and review their answers. The students might need to multiply or add the terms to find different representations.
Supplies	Handout
Mathematical Discourse	Why do we work to combine like terms or rearrange polynomials? Is it an important skill? Why or why not?
Writing/Discussion Prompts	1. What makes some of the modeling problems more challenging than others? 2. Which of the situations you worked through in the homework did you find the most interesting and why?

Name _____ Date _____

Performance Task (continued)

The View Matters

Curriculum Content	
Content Objectives	• Add, subtract, and multiply polynomials. • Identify ways to rewrite an expression. • Factor a polynomial to find the roots of a polynomial equation.
Mathematical Practices	• Construct viable arguments and critique the reasoning of others to determine the best expression for each setting. • Be precise is using the correct unit labels on the solutions.

Rubric

The View Matters	Points	
Answers and reasons may vary but to receive the points, they must be supported. 1. A (the highest point, vertex, will be halfway between the intercepts) Answer: 6 ft 2. C (expression for perimeter has been combined) Answer: 56 m 3. D (the constant term is the height) Answer: 150 ft 4. A (less multiplication without cubes) Answer: 1,332,128 cm^3 5. B (it would be easier to solve for x in the expression) Answer: 1365 yd^2 6. D (expression is simplified so simpler to substitute x) Answer: 306 m^2	18 3 2 1	for each correctly supported answer and correct solution with label for two of the three required components for one of the three components correct
7: Answers vary yet student must answer the questions asked.	6 4 2 1	all questions answered, changes are explained or supported some questions are answered but one question missed or not supported weak answers provided, no evidence of collaboration only a brief statement is correct
Mathematics Practice: Students explain your reasoning on each application.	3	It is clear through the student responses that they recognize the importance mathematics plays in the calculation
Total Points	27 points	

100 Algebra 1
Assessment Book

Copyright © Big Ideas Learning, LLC
All rights reserved.

Name_____ Date_____

Chapter 7 Performance Task (continued)

The View Matters

The way an equation or expression is written can help you interpret and solve problems. Which representation would you rather have when trying to solve for specific information? Why?

For each setting, there is a set of expressions that you might use in each situation. Select the expression that will best help you answer the question. Finish the problem and explain why the selection you made was the best choice.

1. Find the height of the parabolic balloon arch for the prom when the position of the bottom anchors are at $x = 3$ feet and $x = 7$ feet.

   A. $-1.5(x-3)(x-7)$

   B. $\dfrac{318(3-x)(x-7)}{212}$

   C. $-31.5 + 15x - 1.5x^2$

   D. $-\dfrac{3}{2}(x^2 - 10x + 21)$

2. Find the total footage of the fence that surrounds a garden with width $(2x + 3)$ meters and length $(3x - 5)$ meters when $x = 6$.

   A. $2x + 3 + 3x - 5 + 2x + 3x - 5$

   B. $(2x + 3)(3x - 5)$

   C. $10x - 4$

   D. $2(2x + 3) + 2(3x - 5)$

3. You throw a tennis ball straight down from the top of a building with an initial velocity of -30 feet per second. The time (in seconds) after you throw the ball is represented by t. Find the height (in feet) of the building.

   A. $-2(8t^2 + 15t - 75)$

   B. $-8t^2 - 10t + 80 - 8t^2 - 20t + 70$

   C. $-16t^2 - 30(t - 5)$

   D. $-16t^2 - 30t + 150$

Chapter 7 Performance Task (continued)

The View Matters

4. Find the volume of an aquarium with width x centimeters, height $(4x + 5)$ centimeters, and length $(10x - 7)$ centimeters, when the width is 32 centimeters.

    A. $x(4x + 5)(10x - 7)$ **B.** $x + 4x + 5 + 10x - 7$

    C. $6x^3 - x^2 - 15x$ **D.** $15x - 2$

5. Find the area of a farmer's field with width $(3x + 5)$ yards and length $(4x - 1)$ yards, when the width is 35 yards.

    A. $x(12x + 17) - 1$ **B.** $(3x + 5)(4x - 1)$

    C. $12x^2 + 17x - 1$ **D.** $2(3x + 5) + 2(4x + 1)$

6. Find the new area of a rectangular swimming pool when the original length, 20 meters, is decreased by x meters, and the original width, 15 meters, is increased by x meters, when $x = 3$.

    A. $2(20 - x) + 2(15 + x)$ **B.** $300 - x^2$

    C. $(20 - x)(15 + x)$ **D.** $300 + 5x - x^2$

7. Compare your answers with your team or partner. Describe one of the problems where you had different answers. Did either of you change your mind about your answer? Why or why not?

Name_____ Date_____

Performance Task (continued)

Teacher Notes:

Name _____ Date _____

Chapter 8 Quiz
For use after Section 8.3

Identify characteristics of the quadratic function and its graph.

1.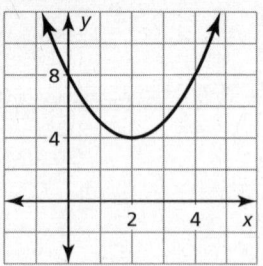

2.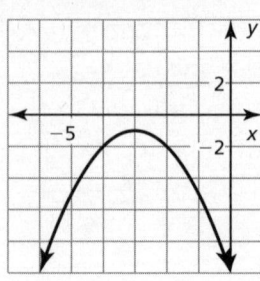

Compare the graph of the function to the graph of $f(x) = x^2$.

3. $h(x) = -2x^2$

4. $p(x) = 4x^2 + 1$

5. $r(x) = 2x^2 - 10$

6. $b(x) = 6x^2$

7. $g(x) = \frac{1}{4}x^2$

8. $m(x) = -\frac{1}{3}x^2 - 4$

Describe the transformation from the graph of f to the graph of g. Write the equation that represents g in terms of x.

9. $f(x) = 4x^2 + 1;\ g(x) = f(x) + 3$

10. $f(x) = \frac{1}{4}x^2 - 5;\ g(x) = f(x) - 3$

Describe the domain and range of the function.

11. $f(x) = -2x^2 - 3x + 9$

12. $y = -4x^2 + 5x + 7$

13. The function $y = -16t^2 + 40$ represents the height y (in feet) of a water droplet t seconds after falling from an icicle.

   a. After how many seconds does the water droplet hit the ground? Round your answer to two decimal places.

   b. A second water droplet falls from a height of 22 feet. Which water droplet hits the ground in the least amount of time? Explain.

14. The function $y = -16t^2 + 50t + 4$ describes the height (in feet) of a discus t seconds after it is released. Describe the domain and range of the function. Find the maximum height of the discus.

Answers

1. _____

_____

_____

2. _____

_____

_____

3. _____

4. _____

5. _____

6. _____

7. _____

8. _____

9. _____

_____

10. _____

_____

11. _____

12. _____

13. a. _____

b. _____

_____

14. _____

_____

_____

_____

Chapter 8 Test A

Graph the function. Compare the graph to the graph of $f(x) = x^2$.

1. $g(x) = 2x^2 - 4$
2. $h(x) = -\frac{1}{3}x^2$
3. $r(x) = 2(x + 3)^2 - 5$

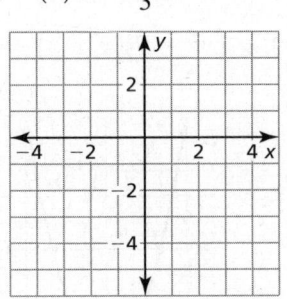
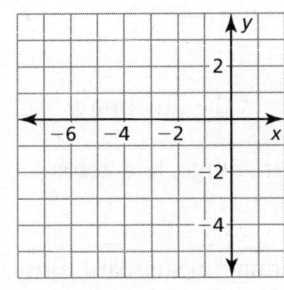

Determine whether the function is *even*, *odd*, or *neither*.

4. $p(x) = 3 - x$
5. $q(x) = -3x^2 - 5x$
6. $r(x) = x^2 + 4$

Use zeros to graph the function.

7. $f(x) = 2x^2 - 8$
8. $h(x) = -(x + 3)(x - 2)$

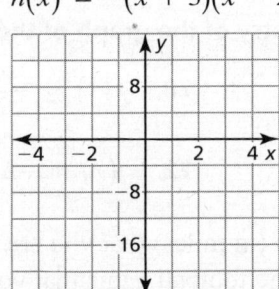

Write a quadratic function in standard form whose graph satisfies the given conditions.

9. passes through $(-5, 0)$, $(1, 0)$, and $(-3, -16)$

10. is even and has a range of $y \leq 2$

11. passes through $(0, 0)$, $(2, 2)$, and $(4, 0)$

Tell whether the table of values represents a *linear*, an *exponential*, or a *quadratic* function. Then write the function.

12.

x	−1	0	1	2	3
y	5	3	1	−1	−3

13.

x	−1	0	1	2	3
y	0.5	1	2	4	8

Answers

1. See left.
2. See left.
3. See left.
4. ___
5. ___
6. ___
7. See left.
8. See left.
9. ___
10. ___
11. ___
12. ___
13. ___

Chapter 8 Test A (continued)

Find the zeros of the function.

14. $f(x) = 3x^2 - 6x - 45$

15. $k(x) = 2(x - 1)(x + 7)$

16. $g(x) = x^3 - 16x$

17. $j(x) = (x^2 - 9)(x + 5)$

18. Consider the graph of the function f.

    a. Find the domain, range, and zeros of the function.

    b. Write the function f in standard form.

    c. Compare the graph of f to the graph of $g(x) = x^2$.

    d. Compare the graph of f to the graph of $h(x) = f(x - 2)$.

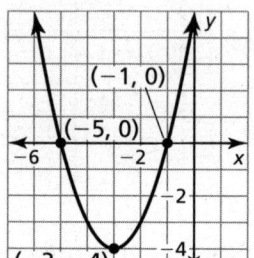

Find the vertex and axis of symmetry of the graph of the function.

19. $y = 2(x - 5)^2 + 1$

20. $f(x) = -\frac{1}{2}(x - 3)^2$

21. $g(x) = 4x^2 - 48x + 7$

22. $j(x) = -2x^2 + 21x - 7$

23. The table shows the distance d (in miles) that you are away from home after driving t minutes away from the football game that you attended that evening.

    a. Tell whether the data can be modeled by a *linear*, an *exponential*, or a *quadratic* function. Explain.

Time, t	10	14	18	22
Distance, d	20	18	16	14

    b. Write a function that models the data.

Plot the points. Tell whether the points appear to represent a *linear*, an *exponential*, or a *quadratic* function.

24. $(-3, 10), (-2, 5), (-1, 2), (0, 1), (1, 2)$

25. $\left(-1, \frac{1}{2}\right), (0, 1), (1, 2), \left(-2, \frac{1}{4}\right), (2, 4)$

26. $(-4, 7), (1, 2), (-3, 6), (-2, 5), (0, 3)$

Answers

14. _____
15. _____
16. _____
17. _____
18. a. _____
    _____
    b. _____
    c. _____
    _____
    d. _____
    _____
19. _____
20. _____
21. _____
22. _____
23. a. _____
    b. _____
24. _____
25. _____
26. _____

Name_____ Date_____

Chapter 8 Test B

Graph the function. Compare the graph to the graph of $f(x) = x^2$.

Answers

1. $g(x) = -\frac{1}{2}x^2 - 4$ 2. $h(x) = 2x^2 + 1$ 3. $r(x) = -(x - 2)^2 - 3$

1. __See left.__

2. __See left.__

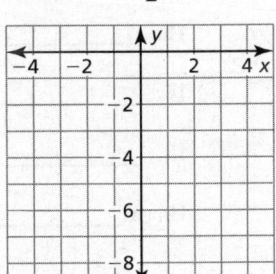 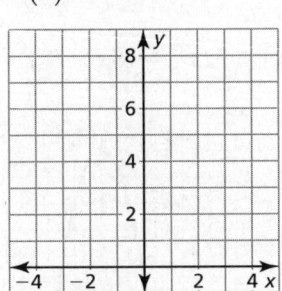

3. __See left.__

4. _____

5. _____

6. _____

Determine whether the function is *even*, *odd*, or *neither*.

7. __See left.__

4. $d(x) = 2x - 3$ 5. $p(x) = -2x^3 - 4x$ 6. $f(x) = 2x^2 + 4x$

8. __See left.__

9. _____

Use zeros to graph the function.

7. $f(x) = -3x^3 + 12x$ 8. $h(x) = -2x^2 - 3x + 5$

10. _____

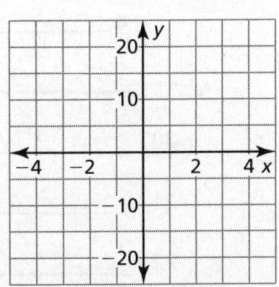 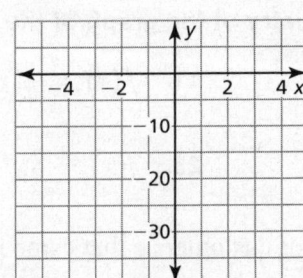

11. _____

12. _____

Tell whether the table of values represents a *linear*, an *exponential*, or a *quadratic* function. Then write the function.

13. _____

14. _____

9.
x	−1	0	1	2	3
y	10	12	14	16	18

10.
x	−2	−1	0	1	2
y	4	2	1	0.5	0.25

11.
x	−1	0	1	2	3
y	−4	−5	−4	−1	4

12.
x	−1	0	1	2	3
y	5	3	5	11	21

Write a quadratic function in standard form whose graph satisfies the given conditions.

13. passes through $(-2, 0)$, $(3, 0)$, and $(2, -2)$

14. is even and has a range of $y \geq -5$

Chapter 8 Test B (continued)

Find the zeros of the function.

15. $f(x) = -2x^2 - 10x + 12$

16. $k(x) = \frac{1}{2}(3x - 1)(2x + 7)$

17. $g(x) = x^4 - 5x^2 + 4$

18. $j(x) = 2x^3 - 18x$

19. Consider the graph of the function f.

    a. Find the domain, range, and zeros of the function.

    b. Write the function f in standard form.

    c. Compare the graph of f to the graph of $g(x) = x^2$.

    d. Compare the graph of f to the graph of $h(x) = -f(x + 3) - 1$.

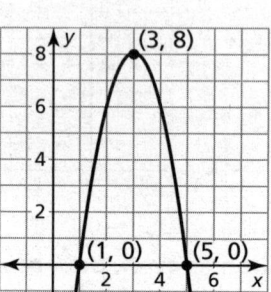

Find the vertex and axis of symmetry of the graph of the function.

20. $y = -3(x + 4)^2 - 2$

21. $f(x) = \frac{1}{4}(x - 1)^2$

22. $g(x) = \frac{1}{2}x^2 - 6x + 10$

23. $j(x) = \frac{2}{3}x^2 + 18x$

24. The table shows the number of customers c that came into a store over a number of hours t.

    a. Tell whether the data can be modeled by a *linear*, an *exponential*, or a *quadratic* function. Explain.

Hours, t	1	2	3	4
Customers, c	3	9	19	33

    b. Write a function that models the data.

Plot the points. Tell whether the points appear to represent a *linear*, an *exponential*, or a *quadratic* function.

25. $(-3, -14), (1, 2), (4, 14), (-2, -10), (0, -2)$

26. $\left(-4, \frac{1}{3}\right), (4, 3), (0, 1), (8, 9)$

Answers

15. _____
16. _____
17. _____
18. _____
19. a. _____
    _____
    _____
    b. _____
    c. _____
    _____
    d. _____
    _____
20. _____
21. _____
22. _____
23. _____
24. a. _____
    b. _____
25. _____
26. _____

Chapter 8 Alternative Assessment

1. Graph the function. Compare the graph to the graph of $f(x) = x^2$.

   a. $g(x) = 3x^2 + 1$

   b. $h(x) = -\frac{1}{3}x^2$

   c. $p(x) = (x-2)^2 - 5$

   d. $q(x) = -\frac{1}{4}(x+3)^2 + 1$

2. Consider the graph of the function f.

   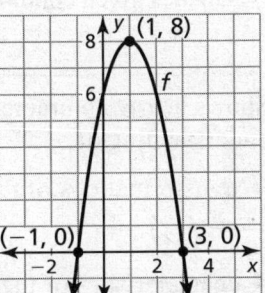

   a. Find the domain, range, and zeros of the function.

   b. Find the vertex and the axis of symmetry of the function.

   c. Write the function f in standard form.

   d. Classify the function as *even*, *odd*, or *neither*.

   e. Compare the graph of f to the graph of $g(x) = x^2$.

   f. Graph $h(x) = f(x+5)$.

3. Write a quadratic function in standard form whose graph satisfies the given conditions. Explain the process you used.

   a. passes through $(-7, 0), (-3, 0),$ and $(-4, 6)$

   b. passes through $(0, 0), (8, 0),$ and $(1, -14)$

   c. is even and has a range of $y \leq 2$

   d. passes through $(3, 0)$ and $(6, 9)$

Name _____ Date _____

Chapter 8 Alternative Assessment Rubric

Score	Conceptual Understanding	Mathematical Skills	Work Habits
4	Completely understands: • transformations of the graph of $f(x) = x^2$ • characteristics of the graph of a quadratic function • writing quadratic function in standard form whose graph satisfies given conditions	Graphs and correctly describes transformations of the graphs of $f(x) = x^2$ Correctly states all the characteristics of the graph of a quadratic function Correctly writes all the quadratic functions in standard form whose graphs satisfy the given conditions	Answers all parts of all four problems All characteristics and graphs are written or drawn carefully and systematically. Work is very neat and well organized.
3	Shows nearly complete understanding of: • transformations of the graph of $f(x) = x^2$ • characteristics of the graph of a quadratic function • writing quadratic function in standard form whose graph satisfies given conditions	Graphs and correctly describes transformations of most of the graphs of $f(x) = x^2$ Correctly states most of the characteristics of the graph of a quadratic function Correctly writes most of the quadratic functions in standard form whose graphs satisfy the given conditions	Answers several parts of all four problems Most characteristics and graphs are written or drawn carefully and systematically Work is neat and organized.
2	Shows some understanding of: • transformations of the graph of $f(x) = x^2$ • characteristics of the graph of a quadratic function • writing quadratic function in standard form whose graph satisfies given conditions	Graphs and correctly describes transformations of some of the graphs of $f(x) = x^2$ Correctly states most of the characteristics of the graph of a quadratic function Correctly writes some of the quadratic functions in standard form whose graphs satisfy the given conditions	Answers some parts of all four problems Characteristics and graphs are written or drawn carefully and systematically Work is not very neat or organized.
1	Shows little understanding of: • transformations of the graph of $f(x) = x^2$ • characteristics of the graph of a quadratic function • writing quadratic function in standard form whose graph satisfies given conditions	Does not answer Exercise 1 Does not state any of the characteristics of the graph of a quadratic function Does not write any of the quadratic functions in standard form whose graphs satisfy the given conditions	Does not attempt any part of any of the problems No characteristics or graphs are written or drawn. Work is sloppy and disorganized.

Name_____ Date_____

 **Performance Task**

Asteroid Aim

Instructional Overview	
Launch Question	Apps take a long time to design and program. One app in development is a game in which players shoot lasers at asteroids. They score points based on the number of hits per shot. The designer wants your feedback. Do you think students will like the game and want to play it? What changes would improve it?
Summary	Grids are provided with asteroids centered at specific coordinates in the xy-plane. Students will be given different rules as they progress through the assessment. Eventually, they will create their own ideas for the scoring. If they create quadratic equations as they learned how to graph in this section, they will be able to hit more asteroids.
Teacher Notes	The students might need help figuring out if they are close enough to the point to call it a hit. Practice with a few equations and decide if an equation hits a point.
Supplies	Handout
Mathematical Discourse	Do you have any app games that require mathematics skills? Would an app need to help students understand how to create equations? Which form of the quadratic equations would be most helpful?
Writing/Discussion Prompts	Which form of a quadratic equation did you find most helpful when trying to hit the point?

Name _____ Date _____

**Chapter 8** **Performance Task** (continued)

Asteroid Aim

Curriculum Content	
Content Objectives	• Create equations in two variables that pass through given points and graph the equations on coordinate axes.
Mathematical Practices	• Be precise in creating equations that hit the asteroids. • Recognize that quadratic equations will have the most potential for hitting more than two points and use the structure of equations to hit the points.

Rubric

Asteroid Aim	Points	
1 Each student would have different equations	**12**	
	4	Two equations written that hit at least one point and scored correctly
	2	Equations written with no hit
2 The student goes back and creates an equation that hits multiple points in one on the first part.	6	A quadratic equation is created that hits at least 3 points.
	4	A quadratic equation is created that hits 2 points.
	2	A linear equation is created that hits two points.
	1	An equation is created that misses.
3 The student provides a set of rules for the game. All the points can be hit by a quadratic equation and an exponential equation.	6	Rules provided, two equations written that hit at least 4 points and scored correctly
	4	Rules provided and two equations hit at least 2 points
	2	The rules would be hard to follow and no quadratic equation is included.
Mathematics Practice: Students make selections that allow them to hit the equations.	3	It is clear through the student responses that they have a better chance of making the shot with different formats.
Total Points	**27 points**	

Name_____ Date_____

 **Performance Task** (continued)

Asteroid Aim

Apps take a long time to design and program. One app in development is a game in which players shoot lasers at asteroids. They score points based on the number of hits per shot. The designer wants your feedback. Do you think students will like the game and want to play it? What changes would improve it?

1. The game starts with a grid and a random set of targets as you see below. You get five points per asteroid hit and two shots. Play the game a few times and record your score. Write your equations and then sketch them on the graph. Record the points you get each round. The app would say you hit the asteroid if you were within 0.3 units of the center.

a.

Shot	Equation	Points
1		
2		

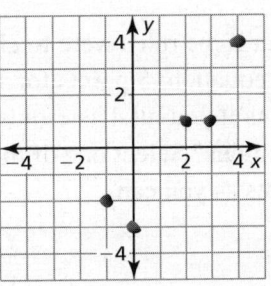

b.

Shot	Equation	Points
1		
2		

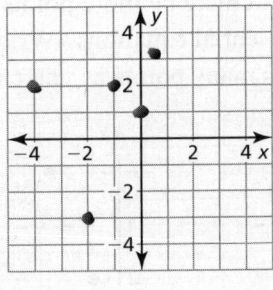

Name _____ Date _____

Performance Task (continued)

Asteroid Aim

c.

Shot	Equation	Points
1		
2		

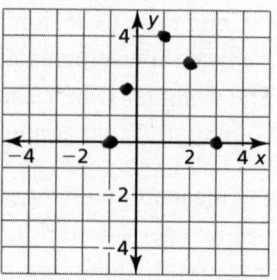

2. What if the rules were to change? What if you earned more points for each hit on one equation: 5 points for 1 hit, 20 points for 2 hits, 45 points for 3 hits, and 80 points if you hit 4 asteroids. Look back at Exercise 1. Would you change your equations? Select one of the three problems and write one equation to get as many points as you can.

3. How would you change the game? What method do you think would be the best? Should there be more points awarded for quadratic equations? What about exponential equations? Write your own rules for your app and then use the grid to get as many points as you can.

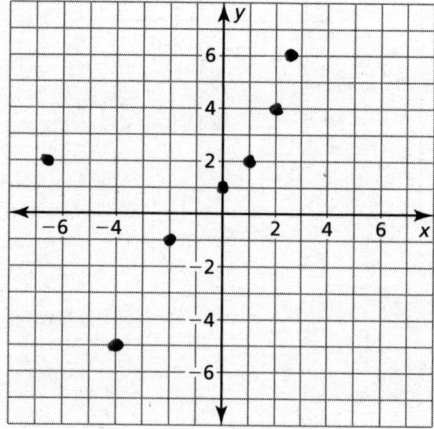

Name_____ Date_____

**Performance Task** (continued)

Teacher Notes:

Name _____ Date _____

Chapters 6–8 Cumulative Test

1. Which expression is different?

   A. $\dfrac{1}{x^{-7}}$
   B. $x^3 \cdot x^4$
   C. $(x^2)^5$
   D. $\dfrac{x^{10}}{x^3}$

2. Evaluate $(2.6 \times 10^2)(3.4 \times 10^{-5})$. Write the answer in scientific notation.

   A. 0.00884
   B. 88.4×10^{-4}
   C. 8.84×10^{-10}
   D. 8.84×10^{-3}

3. Which expression is different?

   A. $81^{1/4}$
   B. 3^0
   C. $\sqrt{9}$
   D. $\sqrt[3]{27}$

4. Simplify $3(3\sqrt{9})^{-2}$.

   A. $\dfrac{1}{3}$
   B. 27
   C. 243
   D. $\dfrac{1}{27}$

5. In 2012, 200,000 people attended the county fair. The attendance increases by 8% per year. Which function represents the attendance t years after 2012?

   A. $y = 200{,}000(1 + 8)^t$
   B. $y = 200{,}000(0.92)^t$
   C. $y = 200{,}000(1.08)^t$
   D. $y = 200{,}000(1.8)^t$

6. Which equation has a solution of $x = 2$?

   A. $2^{x+3} = 2^4$
   B. $4^{x+2} = 4$
   C. $3^{2x-5} = 3^7$
   D. $9^{2x-1} = 9^3$

7. Solve $\left(\dfrac{1}{7}\right)^x = 16{,}807$.

   A. $x = -5$
   B. $x = -2401$
   C. $x = 2401$
   D. $x = 5$

8. What is the 13th term of the geometric sequence where $a_4 = 8$ and $r = 2$?

   A. 26
   B. 4096
   C. 34
   D. 8192

Chapters 6–8 Cumulative Test (continued)

9. Write the explicit rule for the recursive rule $a_1 = -2, a_n = a_{n-1} + 3$.

    A. $a_n = n + 3$
    B. $a_n = 3n - 5$
    C. $a_n = -2n + 3$
    D. $a_n = -2(3)^{n-1}$

10. Which of the following is **not** a polynomial?

    A. $x^3 + 2x^2$
    B. $\sqrt{x^4} - 6x^2 y$
    C. $2x^2 + 3^x$
    D. $\dfrac{1}{x^{-3}} + 5^{-2}$

11. The polynomial expression $(ax^2 + 2)(x^2 - 3x + 1) - (12x^4 - 36x^3)$ is simplified to $14x^2 - 6x + 2$. What is the value of a?

    A. 12
    B. −12
    C. 6
    D. −6

12. What is the value of n in the equation $4n^2 = 20n$?

    A. only 5
    B. only 0
    C. 0 or 5
    D. 0 or −5

13. Which is a factor of the trinomial $x^2 + 6x - 27$?

    A. $x - 19$
    B. $x + 9$
    C. $x - 9$
    D. $x + 3$

14. What value of a would make the graph of $f(x) = ax^2$ open up?

    A. −8
    B. 0
    C. $-\dfrac{1}{2}$
    D. 8

15. How can the graph of $g(x) = x^2 + 4$ be obtained from the graph of $f(x) = x^2$?

    A. by translating $f(x)$ left 4 units
    B. by translating $f(x)$ right 4 units
    C. by translating $f(x)$ up 4 units
    D. by translating $f(x)$ down 4 units

16. Does $f(x) = -3x^2 + 6x - 4$ have a maximum or minimum? Find the value.

    A. maximum; 1
    B. minimum; 1
    C. maximum; −1
    D. minimum; −1

Chapters 6–8 Cumulative Test (continued)

17. Which quadratic function has a vertex of $(-3, 5)$?

A. $y = -3x^2 + 5$

B. $y = 2(x + 3)^2 + 5$

C. $y = -5(x - 5)^2 - 3$

D. $y = \frac{1}{2}(x - 3)^2 + 5$

18. Which quadratic function has zeros of -4 and 7?

A. $f(x) = 2x^2 - 6x - 56$

B. $f(x) = x^2 + 3x - 28$

C. $f(x) = 3(x - 4)(x + 7)$

D. $f(x) = x^2 - 3x + 28$

19. A rectangle has a length of $12x^2y$ and a width of $6xy$.

  a. What is the area of the rectangle?

  b. If the length and width are doubled, what will the area be? Explain your reasoning.

20. Write an expression for the volume of the solid in terms of x.

  a. $V = \pi r^2 h$

  b. $V = \ell wh$

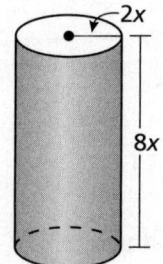
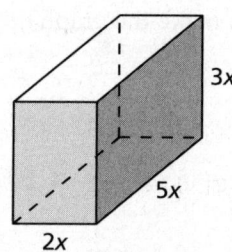

21. The volume of a sphere can be found using the formula $V = \frac{4}{3}\pi r^3$.

  a. Solve the given equation for r.

  b. Find the radius of the sphere whose volume is 288π cubic centimeters.

Chapters 6–8 Cumulative Test (continued)

22. To what power should x be raised on the right side of the equation
$$\frac{x^{-3} \cdot \sqrt[3]{x^2} \cdot x^{2/5}}{x^{-5} \cdot x^0} = x^?\,?$$

23. Which of the following are equal to -4?

| $\sqrt{-16}$ | $\sqrt[3]{-64}$ | $-2^2 \cdot 2^0$ |

| $-\sqrt{-16}$ | $(-8)^{2/3}$ | $\left(\frac{1}{16}\right)^{-1/2}$ |

24. On January 1, you place a penny in your empty piggy bank. On January 2, you place two pennies in your piggy bank. The next day, you place four pennies in your bank. You continue to put money in your bank following this pattern for the month of January.

    a. How many pennies will you add to your bank on January 10?

    b. Is this an arithmetic or geometric sequence?

    c. Write a recursive formula for the sequence.

    d. Write the explicit formula for the sequence.

25. The value of a house is $175,000. It gains 3% of its value every year.

    a. Write a function that represents the value of the house in t years.

    b. Estimate the value of the house in 10 years.

26. Which equation does **not** belong with the other three? Explain your reasoning.

| $y = 200(0.75)^t$ | $y = 100(0.94)^t$ |

| $y = 0.5(0.98)^t$ | $y = 20(1.07)^t$ |

Chapters 6–8 Cumulative Test (continued)

27. Use the sequence 5, 15, 45, 135, ….

    a. Write the next three terms of the sequence.

    b. Is the sequence arithmetic, geometric, or neither? Explain your reasoning.

    c. Write an equation for the nth term of the sequence.

    d. Find a_{12}.

28. Use the sequence 16,384, 4096, 1024, 256, ….

    a. Write a recursive rule for the sequence.

    b. Write the next 3 terms.

29. Which rule does **not** belong with the other three? Explain your reasoning.

$a_n = 2(3)^{n-1}$	$a_1 = 4, a_n = 2a_{n-1}$	$a_n = \frac{1}{2}(7)^{n-1}$	$a_n = -4(5)^{n-1}$

30. A City Council is planning to put a cement walkway around an existing swimming pool. The rectangular pool is twice as long as it is wide, and the walkway will be 3 feet wide on all sides.

    a. Find the area of the pool in terms of its width, w.

    b. Find the area of the walkway and the pool in terms of w.

    c. Find the area of the walkway in terms of w.

    d. Given that the pool is 12 feet wide, what is the area of the walkway?

31. A park has decided to expand its current picnic area as shown in the diagram.

    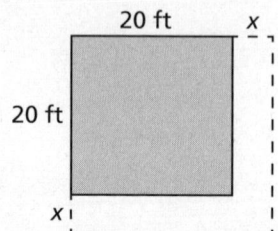

    a. What is the area of the picnic area after the expansion?

    b. Find the area when $x = 5$. What is the area of the extension?

Chapters 6–8 Cumulative Test (continued)

32. A toy rocket is launched upward from a table that is 3 feet tall with an initial velocity of 96 feet per second. The height after t seconds is given by the function $f(t) = -16t^2 + 96t + 3$.

    a. How long will it take the rocket to reach its maximum height?

    b. What is the maximum height?

33. Place each function into one of the 3 categories.

Linear	Exponential	Quadratic

x	0	1	2	3
y	5	7	9	11

x	0	1	2	3
y	3	6	12	24

x	0	1	2	3
y	-3	-2	1	6

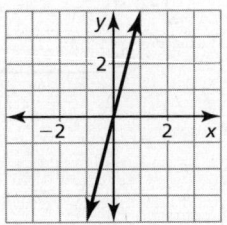

$y = 5^x$

$y = x^2 + 4$

$y = 5x + 3$

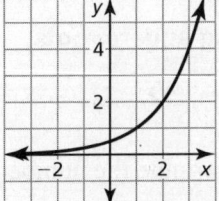

34. Find the dimensions of the rectangle if the area is 23 square inches.

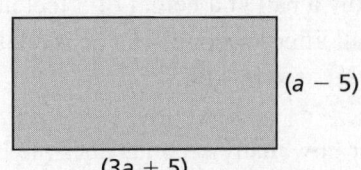

Chapter 9 Quiz
For use after Section 9.3

Simplify the expression.

1. $\sqrt{405x^3}$
2. $\sqrt[3]{-256}$
3. $\dfrac{6}{\sqrt{54}}$
4. $\sqrt[3]{\dfrac{128x^2}{56x^4}}$
5. $\dfrac{4}{3+\sqrt{2}}$
6. $\sqrt{8}\left(5\sqrt{4}-6\sqrt{9}\right)$

Use the graph to solve the equation.

7. $x^2 - 7x + 12 = 0$
8. $x^2 - 4x + 4 = 0$

Solve the equation by graphing.

9. $x^2 + 4x + 20 = 0$
10. $x - 6 = -x^2$

Solve the equation using square roots.

11. $3x^2 = 81$
12. $3x^2 - 5 = 22$
13. $(3x+2)^2 = 81$

14. Explain how to determine the number of real solutions of $x^2 = 50$ without solving.

15. You throw a ball at a height of 5 feet above the ground. The height h (in feet) of the ball after t seconds can be modeled by the equation $h = -16t^2 + 44t + 5$.

   a. After how many seconds does the ball reach a height of 15 feet?

   b. After how many seconds does the ball hit the ground? Round your answer to two decimal places.

Answers

1. _____
2. _____
3. _____
4. _____
5. _____
6. _____
7. _____
8. _____
9. _____See left._____
10. _____See left._____
11. _____
12. _____
13. _____
14. _____
15. a. _____
    b. _____

Chapter 9 Test A

Simplify the expression.

1. $\sqrt{125m^7}$

2. $\sqrt{\dfrac{36x^4}{121}}$

3. $\sqrt{\dfrac{8y^5}{81x^6}}$

4. $\dfrac{4}{\sqrt{6}}$

5. $\dfrac{1}{\sqrt{3}+2}$

6. $3\sqrt{6} - 4\sqrt{24} + 2\sqrt{20}$

Use the graph to solve the equation.

7. $-x^2 - 3x + 4 = 0$

8. $x^2 + 5x = -6$

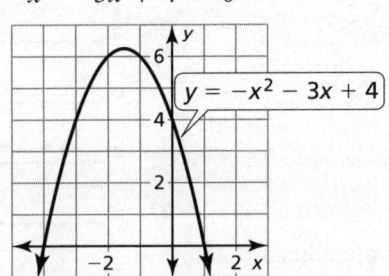
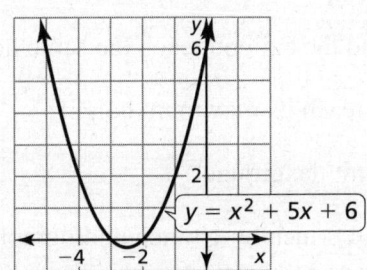

Approximate the zeros of the function. Round to the nearest tenth.

9. $y = x^2 - 7x + 7$

10. $y = -2x^2 - 9x + 10$

Solve the equation by using square roots.

11. $2x^2 + 6 = 14$

12. $(x-1)^2 + 3 = 12$

Solve the equation by completing the square. Round to the nearest tenth, if necessary.

13. $x^2 + 6x = 7$

14. $x^2 - 4x = 10$

Write the function in vertex form by completing the square.

15. $y = x^2 + 2x + 1$

16. $y = -x^2 - 8x - 15$

Solve the equation using any method. Round to the nearest tenth, if necessary.

17. $x^2 - 144 = 0$

18. $x^2 - 6x = -8$

19. $2x^2 - 5x - 13 = 0$

20. $9x^2 + 48x = 36$

Answers

1. _____
2. _____
3. _____
4. _____
5. _____
6. _____
7. _____
8. _____
9. _____
10. _____
11. _____
12. _____
13. _____
14. _____
15. _____
16. _____
17. _____
18. _____
19. _____
20. _____

Name _____ Date _____

Chapter 9 Test A (continued)

Solve the system using any method. Round to the nearest hundredth, if necessary.

Answers

21. $y = x^2 - 3x + 4$
    $y = x + 9$

22. $y = 2x^2 - 4x + 7$
    $y = 9$

23. $y = x^2$
    $y = 1$

24. $y = x^2 - 1$
    $y = \frac{1}{2}x - 3$

21. _____

22. _____

23. _____

24. _____

25. a. _____

25. A ball is thrown upward with an initial velocity of 50 feet per second. The height h (in feet) of the ball after t seconds is given by $h = -16t^2 + 50t$. At the same time, a balloon is rising at a constant rate of 20 feet per second. Its height h in feet after t seconds is given by $h = 20t$.

  a. When do the ball and the balloon reach the same height?

  b. When does the ball reach its maximum height?

  c. When does the ball hit the ground?

b. _____

c. _____

26. a. _____

b. _____

c. _____

27. a. _____

b. _____

26. A picture frame holds an 8-inch by 10-inch photograph. The frame adds a border x inches wide around three sides of the photo. On the fourth side, the frame is wider to accommodate a decoration on the frame. The fourth side is $(3x - 1)$ inches wide, as shown in the figure.

  a. Write a quadratic expression for the combined area of the frame in terms of x.

  b. If the border on the three matching sides is 1 inch, what is the combined area of the frame?

  c. If the combined area of the frame is 165 square inches, find x.

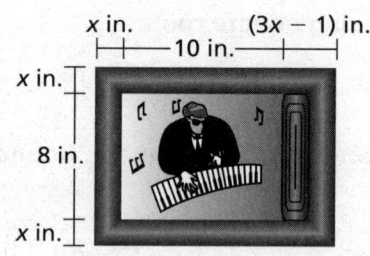

27. You are jumping off the 12-foot diving board at the municipal pool. You bounce up at 6 feet per second and drop to the water. Your height h (in feet) above the water in terms of t seconds is given by $h(t) = -16t^2 + 6t + 12$.

  a. When do you hit the water? Round your answer to the nearest hundredth.

  b. What is your maximum height above the pool?

Chapter 9 Test B

Simplify the expression.

1. $\sqrt[3]{-250x^4y^8}$
2. $\sqrt{\dfrac{12a^3}{36}}$
3. $\sqrt{\dfrac{50y^6}{100x^4}}$
4. $-\dfrac{7}{2\sqrt{3}}$
5. $\dfrac{3}{\sqrt{x}+x}$
6. $2\sqrt{20x^2} - 5x\sqrt{45}$

Use the graph to solve the equation.

7. $5x^3 - 5x = 0$

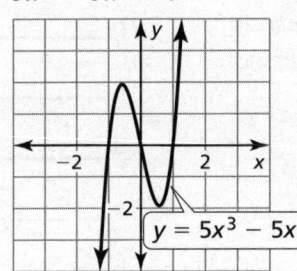

8. $-x^2 - 8x - 6 = 10$

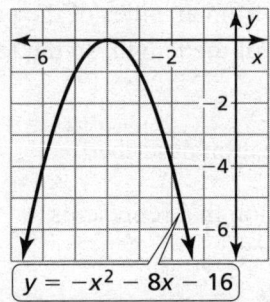

Find the value of the discriminant. Determine the number of real solutions.

9. $x^2 + 14 = 7x$
10. $12x = -4x^2 - 9$
11. $2x^2 - x + 3 = 0$

Solve the equation by using square roots.

12. $\dfrac{2}{3}x^2 - 6 = 0$
13. $-4(x-3)^2 + 1 = 0$

Solve the equation by completing the square. Round to the nearest tenth, if necessary.

14. $13 + x^2 = 10x$
15. $x^2 + 3x = 10$

Write the function in vertex form by completing the square.

16. $y = -x^2 + 4x - 3$
17. $y = x^2 - 5x - 10$

Solve the equation using any method. Round to the nearest tenth, if necessary.

18. $x^2 + 7x = 12 - 4x$
19. $2t^2 + 9t = -7$
20. $80 = 5x^2$
21. $9a^2 = 4 + 7a$

Answers

1. _____
2. _____
3. _____
4. _____
5. _____
6. _____
7. _____
8. _____
9. _____
10. _____
11. _____
12. _____
13. _____
14. _____
15. _____
16. _____
17. _____
18. _____
19. _____
20. _____
21. _____

Name _____ Date _____

Chapter 9 Test B (continued)

Solve the system using any method.

22. $y = x^2 - 7x$
    $y = -10$

23. $y = x^2 - 1$
    $x + 2y = 8$

24. $y = 5 - x^2$
    $y = x + 3$

25. $y = 3x^2 + 4x - 7$
    $2x - y = -1$

Answers

22. _____

23. _____

24. _____

25. _____

26. You are the manager at the town pool. You need to build a fence around the new children's pool area. The pool is 5 feet wide by 8 feet long. You want the fence to be the same distance x from all sides of the pool, except you need to put a lifeguard chair at one end of the length, so the fence should be twice as far away on that side.

    a. Sketch a diagram of the situation described.

    b. Write a polynomial expression that represents the area enclosed by the fence.

    c. If the total area within the fence needs to be 442 square feet, how far away should the fence be from the pool on three of the sides?

    d. How much fencing must you buy?

26. a. _____See left._____

    b. _____

    c. _____

    d. _____

27. a. _____

    b. _____

    c. _____

28. a. _____

    b. _____

27. While vacationing, you decided to go bungee jumping off a cliff into the river. Your height as a function of time is modeled by the function $h(t) = -16t^2 + 16t + 370$, where t is the time in seconds and h is the height in feet.

    a. How long did it take to reach your maximum height?

    b. What is the highest point you reached?

    c. When did you hit the water? Round your answer to the nearest hundredth.

28. You are creating a piece of artwork for the school art show. You need to buy a piece of canvas that is large enough to stretch and secure around a wooden frame. You plan that the length of your piece will be 5 inches more than twice the width, and you will need 1 inch extra on each side to secure the canvas to the frame.

    a. Write a polynomial expression that represents the area of the canvas.

    b. You want the area of the piece to be 364 square inches. How wide should your canvas be? Round your answer to the nearest hundredth.

126 Algebra 1
Assessment Book

Chapter 9 Alternative Assessment

1. Solve the equation using any method. Explain your choice of method.

    a. $x^2 + 3x - 10 = 0$

    b. $-2x^2 - 5x + 9 = 0$

    c. $x^2 - 144 = 0$

    d. $x^2 + 8x = 12$

    e. $7x^2 + x - 6 = 0$

    f. $(5x - 2)^2 = 9$

2. Solve the system using any method.

    a. $y = x^2 + 5x + 3$
       $y = 4x + 5$

    b. $y = -4x^2 + x + 1$
       $y = -4$

    c. $y = \frac{1}{3}(6)^x + 2$
       $y = x^2 + 2x - 5$

3. Consider the quadratic equation $ax^2 + bx + c = 0$. Find the values of a, b, and c so that the graph of its related function has

    a. x-intercepts at $x = 7$ and $x = -2$.

    b. only one x-intercept occurring at $x = 3$.

    c. no x-intercepts and the range incudes only negative values.

    d. one positive x-intercept and one negative x-intercept.

    e. x-intercepts at $x = \sqrt{3}$ and $x = -\sqrt{3}$.

Name _____ Date _____

Chapter 9 Alternative Assessment Rubric

Score	Conceptual Understanding	Mathematical Skills	Work Habits
4	Completely understands: • the methods of solving quadratic equations • the methods of solving a system of nonlinear equations • the relationship between a quadratic equation and its x-intercepts	Uses an appropriate method to solve all of Exercise 1 Correctly solves all of the systems of nonlinear equations Identifies values of a, b, and c for all of Exercise 3	Answers all parts of all problems All equations are written carefully and systematically. Work is very neat and well organized.
3	Shows nearly complete understanding of: • the methods of solving quadratic equations • the methods of solving a system of nonlinear equations • the relationship between a quadratic equation and its x-intercepts	Uses an appropriate method to solve most of Exercise 1 Correctly solves most of the systems of nonlinear equations Identifies values of a, b, and c for most of Exercise 3	Answers several parts of all problems Most equations are written carefully and systematically. Work is neat and organized.
2	Shows some understanding of: • the methods of solving quadratic equations • the methods of solving a system of nonlinear equations • the relationship between a quadratic equation and its x-intercepts	Uses an appropriate method to solve some of Exercise 1 Correctly solves some of the systems of nonlinear equations Identifies values of a, b, and c for some of Exercise 3	Answers some parts of all problems Equations are written carelessly. Work is not very neat or organized.
1	Shows little understanding of: • the methods of solving quadratic equations • the methods of solving a system of nonlinear equations • the relationship between a quadratic equation and its x-intercepts	Does not answer Exercise 1 Does not solve any of the systems of nonlinear equations Identifies no values of a, b, and c for some of Exercise 3	Does not attempt any part of any problem No equations are written. Work is sloppy and disorganized.

Name_____ Date_____

 **Performance Task**

Form Matters

Instructional Overview	
Launch Question	Each form of a quadratic function has its pros and cons. Using one form, you can easily find the vertex, but the zeros are more difficult to find. Using another form, you can easily find the y-intercept, but the vertex is more difficult to find. Which form would you use in different situations? How can you convert one form into another?
Summary	Situations are provided that can be modeled by quadratic equations. The students are then provided with different formats from which they would select their answers. The final problem asks the student to transform an equation to the different formats.
Teacher Notes	The problems throughout the chapter provide good practice for making the selection of different forms of quadratic equations. Students should be encouraged to explain their decisions.
Supplies	Handout
Mathematical Discourse	What is the same about equations that deal with falling objects?
Writing/Discussion Prompts	1. What format would you us to find the vertex of a quadratic equation? What is the most difficult step in the transformation to this form? 2. Provide a situation where one or more of the x-intercepts of the graph of a quadratic equation would be a solution to the equation.

Curriculum Content	
Content Objectives	• Use the method of completing the square to solve quadratic equations and to find the maximum or minimum values of quadratic functions. • Factor and complete the square to find the zeros of a quadratic function. • Create equations in one variable to solve real-life problems.
Mathematical Practices	• Explore the forms of a quadratic equation that provide information for real-life situations.

Chapter 9 Performance Task (continued)

Rubric

Form Matters	Points	
1. a. II; The vertex is visible with the height. b. I; The answer would be found by substitution $t = 3$. c. III; The factored form shows the *x*-intercepts. d. IV; The graph can be used to estimate the solution.	8 2 1	Per part, answer correct with justification One of two parts correct
2. a. III or IV; The answer would be found by substituting and finding the zeros from the factored form or by using the graph. b. II; The vertex is visible and shows the distance from the horizontal to the minimum. c. I or IV; For I, set the equation equal to 3 and solve for *x*, or use the graph and find the intersection with the line $y = -3$.	6 2 1	Per part, answer correct with justification One of two parts correct
3. a. $y = -9.8t^2 + 49t + 58.8$ b. $y = -9.8\left(t - \frac{5}{2}\right)^2 + 120.05$; about 120 m c. $y = -9.8t(t - 6)(t + 1)$; after 6 sec d. $100 = -9.8t^2 + 49t + 58.8$ $0 = -9.8t^2 + 49t - 41.2$ about $t = 1.06964$ sec and 3.93036 sec; using the Quadratic Formula or a graphing calculator e. $y = -9.8t \cdot 2^2 + 49(2) + 58.8 = 117.6$ m f. (graph from -2 to 7, max 130)	12 2 1	Per part, answer correct with justification One of two parts correct
Mathematics Practice: Students select and create the correct forms of equations.	3	Students demonstrate understanding with correct work.
Total Points	**29 points**	

Name_____ Date_____

Chapter 9 Performance Task (continued)

Form Matters

Each form of a quadratic function has its pros and cons. Using one form, you can easily find the vertex, but the zeros are more difficult to find. Using another form, you can easily find the y-intercept, but the vertex is more difficult to find. Which form would you use in different situations? How can you convert one form into another?

For each of the following situations, determine which form of the equation would provide you with the information needed in the most efficient manner. Explain your choice.

1. A football is kicked at ground level with an initial velocity of 64 feet per second.

    I. $y = -16t^2 + 64t$

    II. $y = -16(t - 2)^2 + 64$

    III. $y = -16t(t - 4)$

    IV.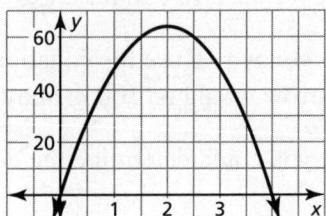

    a. The maximum height of the football is _____.

    b. The height after 3 seconds is _____.

    c. The time when the football hits the ground is _____.

    d. An estimate of the time when the football is 40 feet high is _____.

2. A walking rope bridge from the banks over a small forge can be modeled by a quadratic equation, where x is the distance (in feet) from the left bank and y is the distance (in feet) the bridge sags below the road.

    I. $y = 0.003x^2 - 0.3x$

    II. $y = 0.003(x - 50)^2 - 7.5$

    III. $y = 0.003x(x - 100)$

    IV.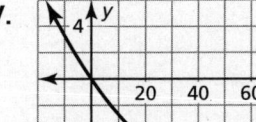

    a. The width of the bridge is _____.

    b. The largest dip (distance from the banks) is _____.

    c. The location(s) on the bridge when a 3-foot-tall child would have her head even with the bank is _____.

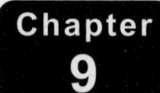

Chapter 9 Performance Task (continued)

3. An arrow is shot straight up from a cliff 58.8 meters above the ground with an initial velocity of 49 meters per second. Let up be the positive direction. Because gravity is the force pulling the arrow down, the initial acceleration of the arrow is −9.8 meters per second squared.

   a. Write the equation that represents this relationship.

   b. Transform the equation to the format that can most easily be used to find the maximum height. What is the height?

   c. Transform the equation to a different format to determine how many seconds it would take for the arrow to hit the ground.

   d. Show how you would use the equation in part (a) to determine approximately when the arrow would be 100 m above the base of the cliff.

   e. Show how you could determine the height of the arrow after 2 seconds.

   f. Sketch the graph of the equation.

Performance Task (continued)

Teacher Notes:

Name _____ Date _____

Chapter 10 Quiz
For use after Section 10.2

Describe the domain of the function.

1. $y = \sqrt{x-6}$
2. $f(x) = 9\sqrt{x}$
3. $y = \sqrt{2-x}$

Compare the graph of the function to the graph of $f(x) = \sqrt{x}$.

4. $f(x) = \sqrt{x} + 7$
5. $n(x) = \sqrt{x-3}$
6. $r(x) = -\sqrt{x-5} + 2$

Compare the graph of the function to the graph of $f(x) = \sqrt[3]{x}$.

7. $b(x) = \sqrt[3]{x+4}$
8. $h(x) = -2\sqrt[3]{x} - 4$
9. $q(x) = \sqrt[3]{-2-x}$

Compare the graphs. Find the value of h, k, or a.

10.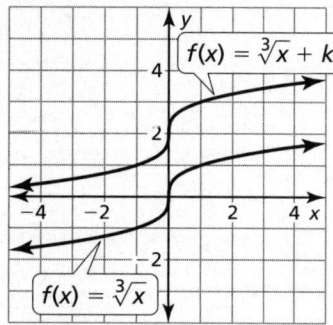

11.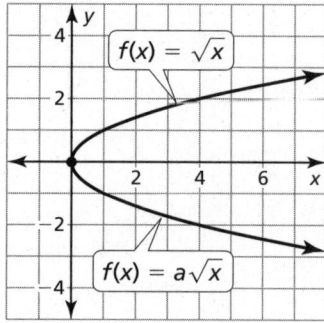

12. The time t (in seconds) it takes a dropped object to fall h feet is given by $t = \tfrac{1}{4}\sqrt{h}$.

    a. Use a graphing calculator to graph the function. Describe the domain and range.

    b. It takes about 13.7 seconds for a ball dropped from the Golden Gate Bridge in San Francisco to reach water below. About how high is the bridge from the water?

13. The formula for velocity is $v = \dfrac{\sqrt{2Em}}{m}$, where E is the kinetic energy (in joules), m is the mass of the object (in kilograms), and v is the velocity of the object (in meters per second). If a basketball with a mass of 0.62 kilogram has 20 joules of kinetic energy, what is its velocity? Round your answer to two decimal places.

Answers

1. _____
2. _____
3. _____
4. _____
5. _____
6. _____
7. _____
8. _____
9. _____
10. _____
11. _____
12. a. _____
    b. _____
13. _____

Chapter 10 Test A

Describe the domain of the function.

1. $f(x) = \sqrt{x+3}$
2. $g(x) = \sqrt{-x} + 2$
3. $h(x) = -\sqrt{x-2} + 1$

Graph the function.

4. $p(x) = \sqrt[3]{x+1} - 1$
5. $n(x) = \sqrt{x} + 3$

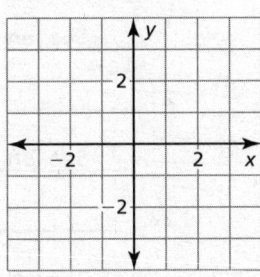
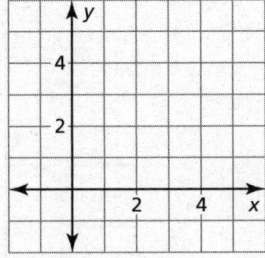

6. $t(x) = \sqrt[3]{x} - 2$
7. $f(x) = -\sqrt{x+1}$

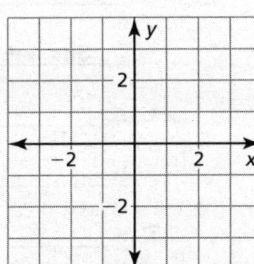
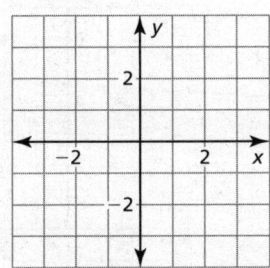

Describe the transformations from the graph of *f* to the graph of *h*.

8. $f(x) = \sqrt{x};\ h(x) = -2\sqrt{x-3} + 4$
9. $f(x) = \sqrt[3]{x};\ h(x) = \frac{1}{2}\sqrt[3]{-x+3} - 1$

Solve the equation. Check your solution.

10. $\sqrt{y} = 4$
11. $2\sqrt{x} - 15 = -7$
12. $2\sqrt{x-1} = 14$
13. $10 - 4\sqrt{2a-7} = -78$
14. $\sqrt[3]{x} + 1 = 5$
15. $\sqrt[3]{2b-7} = -3$
16. $\sqrt{-30 + 11k} = k$
17. $\sqrt{\dfrac{p}{9}} = \sqrt{2p - 170}$

Answers

1. _____
2. _____
3. _____
4. *See left.*
5. *See left.*
6. *See left.*
7. *See left.*
8. _____
9. _____
10. _____
11. _____
12. _____
13. _____
14. _____
15. _____
16. _____
17. _____

Name _____ Date _____

Chapter 10 Test A (continued)

Describe the range of the function.

18. $h(x) = \sqrt{x-5}$

19. $q(x) = -\sqrt{x+3} + 2$

Find the inverse of the function. Then graph the function and its inverse.

20. $f(x) = -2x + 4$

21. $f(x) = 3x + 1$

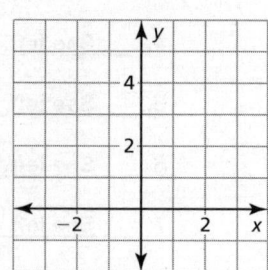 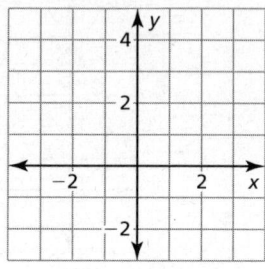

22. $f(x) = x^2 + 4;\ x \geq 0$

23. $f(x) = -2x^2 + 6;\ x \leq 0$

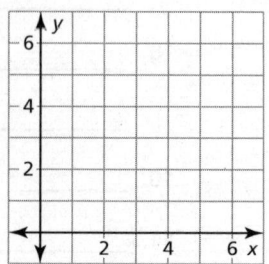 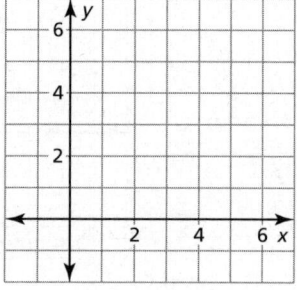

Answers

18. _____

19. _____

20. _____
    See left.

21. _____
    See left.

22. _____
    See left.

23. _____
    See left.

24. a. _____

    b. _____

25. a. _____

    b. _____

24. The velocity (in meters per second) of a car speeding up with time can be modeled by the equation $v(t) = 6\sqrt{t}$.

    a. What is the car's velocity at 4 seconds?

    b. When does the car's velocity reach 24 meters per second?

25. A cylindrical can of soup has a volume of 170 cubic inches. The radius of the soup is found using the formula $r = \sqrt{\dfrac{V}{\pi h}}$, where r is the radius, V is the volume of the can, and h is the height of the can.

    a. If the height of the can is 6 inches, find its radius to the nearest inch.

    b. If the radius of the can is 3 inches, find its height to the nearest inch.

Chapter 10 Test B

Describe the domain of the function.

1. $f(x) = 2\sqrt{x+1}$
2. $g(x) = \sqrt{-x+3}$
3. $h(x) = -2\sqrt{x+4} - 3$

Graph the function.

4. $a(x) = \sqrt[3]{x-2} + 3$

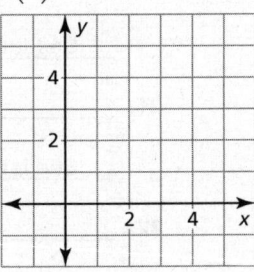

5. $m(x) = \sqrt{-x} + 1$

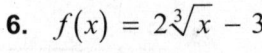

6. $f(x) = 2\sqrt[3]{x} - 3$

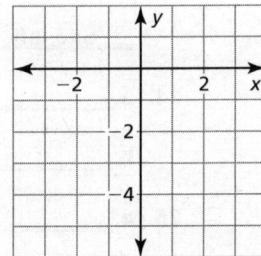

7. $h(x) = -2\sqrt{x-3}$

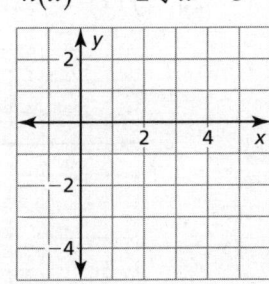

Describe the transformations from the graph of f to the graph of h.

8. $f(x) = \sqrt{x};\ h(x) = \frac{1}{3}\sqrt{x+1} + 5$

9. $f(x) = \sqrt[3]{x};\ h(x) = -2\sqrt[3]{-x+4} + 2$

Solve the equation. Check your solution.

10. $\sqrt{3n+33} = 6$

11. $4\sqrt[3]{\dfrac{x}{2}} = 8$

12. $\sqrt{16-n} = \sqrt{26-2n}$

13. $\sqrt{-56+15r} = r$

14. $v = \sqrt{-6v+3} + 2$

15. $\sqrt{4n-8} = n-2$

16. $63 = 7\sqrt{1-40a}$

17. $2 = \sqrt[3]{-1-m}$

Answers

1. _____
2. _____
3. _____
4. _See left._
5. _See left._
6. _See left._
7. _See left._
8. _____
9. _____
10. _____
11. _____
12. _____
13. _____
14. _____
15. _____
16. _____
17. _____

Chapter 10 Test B (continued)

Describe the range of the function.

18. $h(x) = 2\sqrt{-x-4} + 3$

19. $q(x) = -\frac{1}{2}\sqrt{x-1} - 5$

Find the inverse of the function. Then graph the function and its inverse.

20. $f(x) = -x + 5$

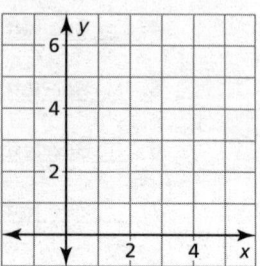

21. $f(x) = 4x + 2$

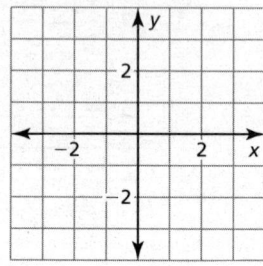

22. $f(x) = x^2 - 3$; $x \leq 0$

23. $f(x) = \frac{1}{2}x^2 - 4$; $x \geq 0$

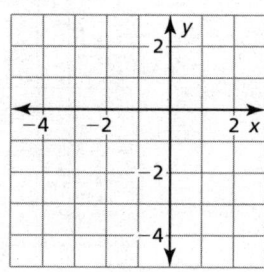

Answers

18. _____

19. _____

20. _____

   See left.

21. _____

   See left.

22. _____

   See left.

23. _____

   See left.

24. a. _____

    b. _____

25. a. _____

    b. _____

24. The amount of time t (in seconds) it takes a pendulum to complete a full swing is given by $t = 2\pi\sqrt{\dfrac{\ell}{32}}$, where ℓ is the length of the pendulum's arm in feet.

   a. A giant swing completes a full swing in 7 seconds. How long is the arm? Give your answer to two decimal places.

   b. If the length of the arm is decreased, what happens to the amount of time it takes for a full swing?

25. A regulation basketball has a volume of 456 cubic inches. The radius of the ball is found using the formula $r = \sqrt[3]{\dfrac{3V}{4\pi}}$, where r is the radius and V is the volume of the ball.

   a. If the volume of the ball is 400 cubic inches, what is the radius of the ball? Give your answer to two decimal places.

   b. If the radius of the ball is 4.5 inches, is the ball at regulation volume?

Chapter 10 Alternative Assessment

1. Graph the function f. Describe the domain and range. Compare the graph of f to the graph of g.

   a. $f(x) = -\sqrt{x-4}$; $g(x) = \sqrt{x}$

   b. $f(x) = 2\sqrt{x} - 5$; $g(x) = \sqrt{x}$

   c. $f(x) = \sqrt[3]{x-2} + 4$; $g(x) = \sqrt[3]{x}$

   d. $f(x) = -\frac{1}{2}\sqrt[3]{x-1} - 3$; $g(x) = \sqrt[3]{x}$

2. Solve the equation. Check your solution(s).

   a. $4 + \sqrt{x} = 9$

   b. $\sqrt{4x+1} - 5 = 1$

   c. $\sqrt{-6x-9} = \sqrt{11+4x}$

   d. $x - 5 = 2\sqrt{x-5}$

   e. $\sqrt{4x+2} = \sqrt{4x-3}$

3. Find the inverse of the function. State the domain and range of the inverse function.

   a. $f(x) = 7x - 2$

   b. $f(x) = \frac{1}{3}\sqrt{x-4} + 2$

   c. $f(x) = -\frac{1}{2}(x+3)^2$, $x \geq -3$

   d. $f(x) = \sqrt[3]{x} - 4$

   e. $f(x) = \frac{1}{4}x^2 - 1$, $x \leq 0$

Name_____ Date _____

Chapter 10 Alternative Assessment Rubric

Score	Conceptual Understanding	Mathematical Skills	Work Habits
4	Completely understands: • graphs of square root and cube root functions • solving radical equations • inverse of a function	Correctly graphs all of the functions and describes the domain and range Correctly solves all of the radical equations Correctly finds all of the inverses and describes the domain and range	Answers all parts of all of the problems All equations and graphs are written or drawn carefully and systematically. Work is very neat and well organized.
3	Shows nearly complete understanding of: • graphs of square root and cube root functions • solving radical equations • inverse of a function	Correctly graphs most of the functions and describes the domain and range Correctly solves most of the radical equations Correctly finds most of the inverses and describes the domain and range	Answers several parts of all of the problems Most equations and graphs are written or drawn carefully and systematically. Work is neat and organized.
2	Shows some understanding of: • graphs of square root and cube root functions • solving radical equations • inverse of a function	Correctly graphs some of the functions Correctly solves some of the radical equations Correctly finds some of the inverses	Answers some parts of all of the problems Equations and graphs are written or drawn carelessly. Work is not very neat or organized.
1	Shows little understanding of: • graphs of square root and cube root functions • solving radical equations • inverse of a function	Does not graph any of the functions Does not solve any of the radical equations Does not find any of the inverses	Does not attempt any part of the problems No equations or graphs are written or drawn. Work is sloppy and disorganized.

Name_____ Date_____

Performance Task

Medication and the Mosteller Formula

Instructional Overview	
Launch Question	When taking medication, it is critical to take the correct dosage. For children in particular, body surface area (BSA) is a key component in calculating that dosage. The Mosteller Formula is commonly used to approximate body surface area. How will you use this formula to calculate BSA for the optimum dosage?
Summary	This task is designed to help students see another application of radical equations and work with multiple measurement units.
Teacher Notes	Children's medicine doses are often calculated from adult doses by using age, body weight, and/or body surface area. The most reliable methods are based on body surface area. More information can be found at http://www.halls.md/body-surface-area/refs.htm, and there is a calculator at http://www.halls.md.body-surface-area/bsa.htm.
Supplies	Handout
Mathematical Discourse	Why does it matter that a person gets the correct dosage of medication?
Writing/Discussion Prompts	1. If you were not going to use a formula to estimate your body surface area, how would you figure it out? 2. What would be your dose of a medication if your BSA was 2.13 square meters and the dose per square meter was 1 teaspoon?

Name _____ Date _____

Chapter 10 Performance Task (continued)

Medication and the Mosteller Formula

Curriculum Content	
Content Objectives	• Use radical equations to solve real-life problems.
Mathematical Practices	• Explore the use of mathematics in medicine.

Rubric

Medication and the Mosteller Formula	Points	
1. a. $H = 163$ cm, $W = 63.5$ kg $\sqrt{\dfrac{63.5 \cdot 163}{3600}} \approx 1.7$ m^2 b. *Sample answer:* 10.9 kg and 84 cm c. Answer will vary depending on the student; you can check it with the calculator given in the teacher notes.	9 3 1	Correct work and solution on each section Minor errors
2. Use the relationships 1 kg = 2.2046226218 lb and 1 in. = 2.54 cm.	4 2	Substitutions are correct and calculations equate the equations. Most of the work is correct with minor errors.
3. Graph is correct. (graph showing square-root-like curve with y-axis labeled to 2, x-axis labeled 20, 40, 60, 80)	4 2	The graph is correct. Minor errors
4. It could be approximated by using a graph or by substituting values into the equation. Answer: 180 lb	4	Substitutions are correct and calculations equate the equations.
Mathematics Practice: The students are able to work with the equations and correct units.	3	Students demonstrate understanding with correct work.
Total Points	**24 points**	

Chapter 10 Performance Task (continued)

Medication and the Mosteller Formula

When taking medication, it is critical to take the correct dosage. For children in particular, body surface area (BSA) is a key component in calculating that dosage. The Mosteller Formula is commonly used to approximate body surface area. How will you use this formula to calculate BSA for the optimum dosage?

1. The Mosteller formula is $BSA = \sqrt{\dfrac{H \cdot W}{3600}}$, where BSA is measured in square meters, H is the height (in centimeters), and W is the weight (in kilograms).

   a. Find the BSA for an adult weighing 63.5 kilograms and with a height of 163 centimeters.

   b. The average BSA for a 2-year old child is approximately 0.5 square meter. What would be a possible height and weight that would produce this BSA?

   c. Convert your weight to kilograms and your height to centimeters and calculate your BSA.

2. $BSA\ (m^2) = ([\text{Height (in.)} \times \text{Weight (lb)}]/3131)^{1/2}$ is the same formula using inches and pounds. Use the conversion factors you found when converting your own height and weight and show that these are the same formulas.

3. Given the Mosteller formula from Exercise 2 and a height of 5 feet 10 inches, graph the equation for the BSA dependent on the weight.

4. Estimate the approximate weight that would yield a BSA of 2 square meters.

Chapter 11 Quiz
For use after Section 11.3

Find the mean, median, and mode of the data set. Which measure of center best represents the data? Explain.

1. **Hours Spent Practicing**

8	$6\frac{3}{4}$	5
$5\frac{1}{2}$	4	$7\frac{1}{2}$

2. **Building Height (feet)**

1250	1525	2175
1400	1400	1650
1550	1350	1400

Find the range and standard deviation of each data set. Then compare your results.

3. Number of students in a class
   Female: 12, 15, 11, 12, 17
   Male: 10, 15, 14, 8, 11

4. Number of goals scored in a year
   Juniors: 16, 20, 4, 7, 10, 8, 7, 10
   Seniors: 20, 24, 12, 4, 8, 15, 12, 10

5. The table shows the prices of eight electronic tablets.

Price (dollars)	188	480	250	151	310	200	295	190

   a. Find the mean, median, mode, range, and standard deviation of the prices.

   b. Identify the outlier. How will it affect the mean, median, and mode?

   c. Make a box-and-whisker plot that represents the data. Find and interpret the interquartile range of the data. Identify the shape of the distribution.

6. The table shows the times it takes to cook 15 meals.

Time (minutes)				
25	20	27	18	40
65	50	35	35	40
40	58	90	38	20

   a. Display the data in a histogram using four intervals beginning with 11–30.

   b. Which measures of center and variation best represent the data?

Answers

1. _____
   _____
   _____
   _____

2. _____
   _____
   _____
   _____

3. _____
   _____

4. _____
   _____

5. a. _____
   b. _____
   c. _See left._

6. a. _See left._
   b. _____

Name_____ Date_____

Chapter 11 Test A

Find the mean, median, and mode of the data set. Which measure of center best represents the data?

1. 2, 3, 3, 3, 6, 7, 5, 4, 5, 2
2. 10, 11, 15, 12, 11, 13, 13, 10, 9, 11

Use the box-and-whisker plot to find the given measure.

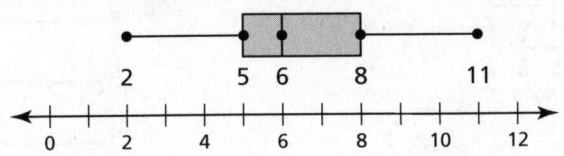

3. least value
4. first quartile
5. median
6. range
7. greatest value
8. third quartile

Make a box-and whisker plot that represents the data.

9. 11, 8, 4, 8, 5, 7, 9, 10, 8, 9, 8
10. 24, 33, 56, 42, 35, 45, 33

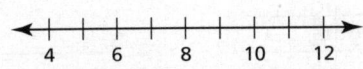

11. The table shows the hourly wages of eight high school seniors in Massachusetts.

    a. Identify the outlier. How does the outlier affect the mean, median, and mode?

    b. Describe one potential explanation for the outlier.

Wages (dollars)			
8.10	9.00	8.50	10.00
9.50	8.25	9.15	17.00

Find the range and standard deviation of each data set. Then compare your results.

12. Students' test scores
    Science: 86, 85, 90, 75, 56, 80
    Math: 75, 73, 93, 84, 93, 70

13. Males' heights
    Freshmen: 64, 62, 56, 63, 72
    Seniors: 72, 68, 67, 60, 70

Tell whether the data are *qualitative* or *quantitative*.

14. price of women's shoes
15. brands of sneakers at store
16. cost of monthly electric bill
17. breeds of dogs at a competition

Answers

1. _____
   _____
2. _____
   _____
3. _____
4. _____
5. _____
6. _____
7. _____
8. _____
9. See left.
10. See left.
11. a. _____
    b. _____
12. _____
    _____
    _____
13. _____
    _____
    _____
14. _____
15. _____
16. _____
17. _____

Name _____ Date _____

Chapter 11 Test A (continued)

Find the value of x.

18. 8.5, 9, 7.5, 6, 7, x; The mean is 8.

19. 1, 3, 3, 5, 2, 4, x; The mean is 3.

20. The double box-and-whisker plot represents the monthly profit for one year for two companies.

   a. Identify the shape of each distribution.

   b. Which company's profit is more consistent on a monthly basis? Explain.

   Company B: 4 5 6 7 11
   Company A: 4 5 7 8 9
   Profit (millions of dollars)

   c. Which company has the single best month during the year?

21. The table shows the heights (in inches) of the 2012 NBA All-Star Game players.

Players' Height (in.)											
70	74	83	85	70	75	76	83	83	81	86	78
79	75	82	75	79	73	75	81	82	78	72	84

   a. Display the data in a histogram using five intervals beginning with 69–72.

   b. Which measure of center best represents the data? Explain.

   Players' Heights (histogram with Frequency axis showing 0, 4, 8; Height (inches) axis)

Answers

18. _____

19. _____

20. a. _____

   b. _____

   c. _____

21. a. ___See left.___

   b. _____

22. _____

23. _____

24. a. ___See left.___

   b. _____

Choose an appropriate data display for the situation. Explain your reasoning.

22. price of a home over 10 years

23. different types of pets in 100 homes

24. You conduct a survey that asks 350 students in your class about whether they prefer pizza or chicken for school lunch. One hundred eighty-nine males respond, 95 of which prefer pizza. Fifty-two females prefer the chicken.

   a. Organize the results in the two-way table provided.

   b. What percent of females like the pizza?

	Male	Female	Total
Pizza			
Chicken			
Total			

Name_____ Date_____

Chapter 11 Test B

Find the mean, median, and mode of the data set. Which measure of center best represents the data?

Answers

1. 1, 8, 7, 3, 6, 7, 5, 10, 5, 4
2. 12, 20, 13, 10, 12, 12, 14, 15, 18, 19

1. _____

Use the box-and-whisker plot to find the given measure.

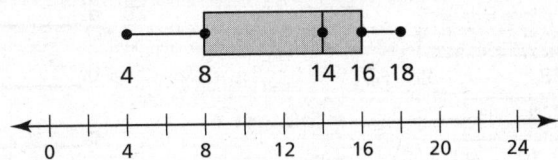

2. _____

3. _____

4. _____

3. least value
4. first quartile
5. median

5. _____

6. range
7. greatest value
8. third quartile

6. _____

Make a box-and whisker plot that represents the data.

7. _____

9. 30, 36, 40, 28, 32, 48, 42
10. 10, 9, 7, 8, 3, 7, 9, 2, 8, 7, 9

8. _____

9. ___See left.___

10. ___See left.___

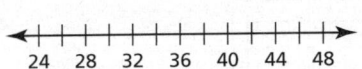 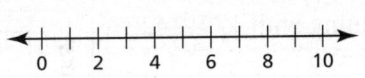

11. a. _____

b. _____

11. The table shows the test scores for eight honors geometry students on an exam.

12. _____

a. Identify the outlier. How does the outlier affect the mean, median, and mode?

Test Scores			
45	78	83	85
84	90	80	86

b. Describe one potential explanation for the outlier.

13. _____

Find the range and standard deviation of each data set. Then compare your results.

12. Highest test scores by class
Period 2: 92, 95, 87, 90, 92, 93, 88
Period 4: 100, 88, 83, 89, 92, 91, 85

13. Height of 18-year old adults (in.)
Male: 70, 67, 72, 65, 69, 71, 73
Female: 64, 62, 65, 63, 67, 61, 70

14. _____

15. _____

16. _____

Tell whether the data are *qualitative* or *quantitative*.

17. _____

14. brands of athletic clothing
15. nations competing in the Olympics

16. monthly snowfall in inches
17. average temperatures in a city

Copyright © Big Ideas Learning, LLC
All rights reserved.

Algebra 1 147
Assessment Book

Chapter 11 Test B (continued)

Find the value of x.

18. 2, 4, 7, 0, 2, 3, x; The mean is 3.

19. 17, 15, 16, 18, x; The mean is 17.

20. The double box-and-whisker plot represents the quiz scores of two students over the course of the year in their algebra class.

   a. Identify the shape of each distribution.

   b. Which student was more consistent on his or her quiz scores? Explain.

   c. Which student has the single best quiz score?

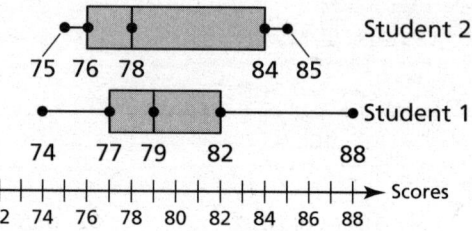

21. The table shows the weight of several sumo wrestlers in kilograms.

Wrestler's Weight (kg)									
175	150	134	180	143	127	159	157	151	203
143	134	135	150	132	169	129	135	139	153

   a. Display the data in a histogram using five intervals beginning with 127–142.

   b. Which measure of center best represents the data? Explain.

22. You conduct a survey that asks 350 students in your class about whether they plan to go to prom this year. One hundred seventy-one males respond, 100 of which say that they plan to go. Ninety females say they are not planning to attend.

   a. Organize the results in the two-way table provided.

   b. If no student changes his or her mind between now and the prom, what percent of the prom attendees will be female?

	Male	Female	Total
Yes			
No			
Total			

Answers

18. _____

19. _____

20. a. _____

    b. _____

    c. _____

21. a. _See left._

    b. _____

22. a. _See left._

    b. _____

Alternative Assessment

1. Determine which measures of center and variation best represent the data.

   a. The data is skewed left.

   b. In a box-and-whisker plot, the length of the box to the left of the median and the length of the box to the right of the median are the same.

   c. A stem-and-leaf plot has nine stems and each has six or seven leaves.

   d. The data is skewed right.

   e. The data is symmetric.

   f. A histogram has six bars. The bars on the left are higher and the bars on the right are gradually lower.

2. The table shows the weights of 10 randomly selected cats.

Weights of Cats (pounds)				
9.3	10.2	8.9	7.5	9.5
9.6	11.0	9.6	10.4	9.8

   a. Find the mean, median, mode, range, and standard deviation of the weights.

   b. Repeat part (a) when all the cats have gained 0.4 pound.

   c. Which data display best represents the data, a stem-and-leaf plot or a box-and-whisker plot? Explain.

3. You conduct a survey by asking 158 students in your class whether they prefer running on a treadmill or running outside. Ninety-one females respond, 64 of which prefer to run on a treadmill. Fifty-two males prefer running outside.

   a. Organize the results in a two-way table. Find and interpret the marginal frequencies.

   b. What percent of males prefer to run on a treadmill?

Name _____ Date _____

Chapter 11 Alternative Assessment Rubric

Score	Conceptual Understanding	Mathematical Skills	Work Habits
4	Completely understands: • measures of center and variation • box-and-whisker plots • two-way tables	Correctly determines which measures of center and variation best represent all the data Correctly finds all the five-point summary and quartiles Correctly organizes and interprets all the data in a two-way table	Answers all parts of the three problems All analyses and displays are written or drawn carefully and systematically. Work is very neat and well organized.
3	Shows nearly complete understanding of: • measures of center and variation • box-and-whisker plots • two-way tables	Correctly determines which measures of center and variation best represent most of the data Correctly finds most of the five-point summary and quartiles Correctly organizes and interprets most of the data in a two-way table	Answers several parts of the three problems Most analyses and displays are written or drawn carefully and systematically. Work is neat and organized.
2	Shows some understanding of: • measures of center and variation • box-and-whisker plots • two-way tables	Correctly determines which measures of center and variation best represent some of the data Correctly finds some of the five-point summary and quartiles Correctly organizes and interprets some of the data in a two-way table	Answers some parts of the three problems Analyses and displays are written or drawn carelessly. Work is not very neat or organized.
1	Shows little understanding of: • measures of center and variation • box-and-whisker plots • two-way tables	Does not determine which measures of center and variation best represent the data Does not find the five-point summary or the quartiles Does not organize or interpret the data in a two-way table	Does not attempt any parts of the three problems No analyses and displays are written or drawn. Work is sloppy and disorganized.

Name_____ Date_____

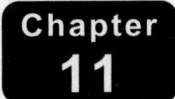 **Performance Task**

College Student Study Time

Instructional Overview	
Launch Question	Data from a small survey at a state university could provide insight into the amount of study time necessary to be successful in college. Based on the information you find when you organize the data, what advice should you give your peers? How will you support your conclusions?
Summary	Students are provided with some data about study hours and GPA. They are asked to create graphs from the chapter including one of their own choosing.
Teacher Notes	The data is nearly linear and should imply that it has some correlation.
Supplies	Handout
Mathematical Discourse	Do you believe you will study more or less when you go to college? Why?
Writing/Discussion Prompts	1. Are students with better GPAs smarter or do they work harder? Support your opinion. 2. What types of graphs provide the best information? Does it depend on the data? Why or why not?

Curriculum Content	
Content Objectives	• Find the mean, median, and mode of data sets. • Use and interpret a box-and-whisker plot. • Choose and create an appropriate data display to analyze data.
Mathematical Practices	• Summarize data to make a conclusion about how the number of hours studied affects GPA.

Chapter 11 Performance Task (continued)

Rubric

College Student Study Time	Points	
1. mean = 20.53; median = 21; mode = 22	3	Correct work and solution for each measure
	1	minor errors
2. mean = 2.83; median = 2.9; modes = 2.5, 3.9	3	Correct work and solution for each measure
	1	minor errors
3. min = 6; Q1 = 15; med = 21; Q3 = 25; max = 42 *Sample answer:* The data is reasonably symmetric, except for an outlier at 42 hours that skews the distribution to the right.	6	The five-number summary is correct and the plot is correct.
	1	per correct part
4. *Sample answer:* a stem-and-leaf plot because you see a picture while maintaining the data	4	The student makes a choice, creates it correctly, and supports the choice.
5.	3	The student correctly plots the points and discusses the shape and outlier.
	1	Only some of the points are plotted or there are multiple errors in the plots.
6. *Sample answer:* There appears to be a strong positive correlation, which suggests that the more you study, the higher your GPA.	4	A reasonable summation is provided.
Mathematics Practice: The students are able to work with the equations and correct units.	3	The student demonstrates understanding with correct work.
Total Points	**26 points**	

Name_____ Date_____

Performance Task (continued)

College Student Study Time

Data from a small survey at a state university could provide insight into the amount of study time necessary to be successful in college. Based on the information you find when you organize the data, what advice should you give your peers? How will you support your conclusions?

Here is the data collected. The first number is the number of hours the students studied in an average week and the second number is that student's GPA for the semester.

Hours	10	12	15	6	24	30	22	25	42	21	22	18	28	16	17
GPA	2.0	1.9	2.4	1.4	3.7	3.9	3.3	3.5	3.9	2.9	3.0	2.5	3.4	2.1	2.5

1. Find the three measures of center for the hours the students studied.

2. Find the three measures of center for the GPA.

3. Make a box-and-whisker plot to represent the number of hours studied. Label all key components. Discuss if the data is skewed or not. Is there an outlier?

4. What is an additional graph that you could make that you think would best represent the hours studied. Support your choice.

5. Plot the points as ordered pairs. What type of trend do you see in the data?

6. If you were to summarize what this data tells you about studying in college, what would you say?

Chapters 9–11 Cumulative Test

1. Which set of data has the same median and mean?

   A. 35, 56, 34, 44, 52, 12, 34, 45

   B. 24, 34, 32, 16, 45, 38, 28

   C. 32, 23, 22, 33, 33, 23, 32, 23, 22, 32

   D. 86, 24, 65, 65, 24, 24

2. The x-intercepts of the graph of a quadratic function are given by $x = \dfrac{7 \pm \sqrt{73}}{4}$. Which of the following could be the quadratic equation represented by the related function?

   A. $2x^2 - 7x - 13 = -10$

   B. $2x^2 + 7x - 3 = 0$

   C. $-2x^2 + 7x - 3 = 0$

   D. $2x^2 + 7x - 13 = -10$

3. Which of the following has an extraneous solution?

   A. $\sqrt{x+2} = x$ B. $\sqrt{x} + 4 = 8$ C. $\sqrt{5x+1} = 6$ D. $\sqrt{x-3} = 4$

4. Which of the following is **not** part of a transformation from the graph $f(x) = \sqrt{x}$ to the graph $h(x) = 2\sqrt{x-3} + 4$?

   A. 3 units right

   B. 4 units up

   C. stretch by a factor of 2

   D. 3 units left

5. What are the roots of the quadratic function below?

   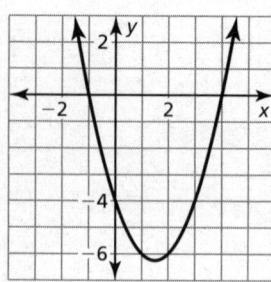

   A. -4

   B. -1 and 4

   C. -4 and -1

   D. $-4, -1,$ and 4

6. Which value of x makes $3\sqrt{13x} = 21\sqrt{13}$?

   A. 7

   B. 14

   C. 49

   D. 98

Chapters 9–11 Cumulative Test (continued)

7. Which ordered pairs would be elements of the inverse of the function $\{(-2, 3), (4, 5), (0, 0), (7, -2), (8, 1)\}$?

(−3, 2)	(−4, −5)	(−2, 7)	(0, 0)
(8, −1)	(3, −2)	(1, 8)	(5, 4)

8. The height of an equilateral triangle can be found using the formula $h = \dfrac{\sqrt{3}}{2}a$, where a is the side length of the equilateral triangle.

  a. If the side length of the equilateral triangle is 5 centimeters, what is the exact height of the triangle?

  b. If the height of the equilateral triangle is 6 inches, what is the perimeter of the triangle?

9. You keep track of the number of algebra problems your teacher assigns for 3 weeks.

Number of Problems Assigned				
20	15	25	12	0
13	17	9	21	11
25	14	16	19	5

  a. Display the data in a histogram using 6 intervals beginning with 0–4.

  b. Describe the shape of the distribution.

  c. How would you interpret the data?

Name _____ Date _____

Chapters 9–11 Cumulative Test (continued)

10. The speed that a tsunami can travel is modeled by $S = 356\sqrt{d}$, where S is the speed (in kilometers per hour) and d is the average depth of the water in kilometers.

    a. What is the speed of the tsunami when the average water depth is 0.4 kilometer?

    b. Solve the equation for d.

    c. A tsunami is traveling at 600 kilometers per hour. What is the average depth of the water?

11. A right triangle has a base that is 8 centimeters longer than its height. If the area of the triangle is 93 square centimeters, what is the height?

12. Consider the parabola $y = x^2 + 14x - 51$.

    a. Write the equation of the parabola in vertex form.

    b. Is the vertex a maximum or minimum? State the vertex.

    c. Find the zeros of the function.

13. An object is launched directly upward at 64 feet per second from a platform 80 feet high so that $h(t) = -16t^2 + 64t + 80$.

    a. What will be the object's maximum height?

    b. When will it obtain this height?

    c. How long will it take it to reach the ground?

14. Which of the following are solutions to the system $\begin{cases} x + y = 1 \\ y = x^2 - 5 \end{cases}$?

    | (−3, 2) | (2, −1) | (2, −3) |

    | (−3, 4) | (4, −3) | (−1, 2) |

Chapters 9–11 Cumulative Test (continued)

15. The side of a square is given by $s = \sqrt{A}$, where A is the area of the square.

    a. Describe the domain of the function.

    b. What is the minimum value of the function?

    c. Does the minimum value make sense in this situation?

    d. Explain why there is no maximum value.

16. The volume of a sphere can be found using the formula $V = \frac{4}{3}\pi r^3$.

    a. Solve this equation for r.

    b. A regulation size basketball has a volume of 455.9 cubic inches. What is the diameter of the basketball?

    c. If the circumference of the basketball is 29 inches, is it a regulation basketball?

17. Gary is in a car at the top of a hill of a roller coaster. The distance d (in feet) of the car from the ground as the car descends is given by the equation $d = 400 - 16t^2$, where t is the number of seconds it takes the car to travel down the hill.

    a. Find and interpret the x-intercepts of the graph of the equation.

    b. How long would it take the car to descend the hill if it were 144 feet tall?

18. Which equation does not belong with the other three? Explain your reasoning.

$x^2 = 21$	$-6x^2 = -414$
$2x^2 = -144$	$x^2 + 7 = 88$

Chapters 9–11 Cumulative Test (continued)

19. The area A of a square with side length s is given by the formula $A = s^2$.

    a. Solve the formula for s.

    b. Explain why the $\pm$ sign is not necessary when finding the side length using the derived formula in part (a).

    c. Use the formula from part (a) to find the side length of a square whose area is $50x^2$ square inches.

20. A teacher surveyed his classes about their favorite choice for school lunches. The results of the survey are shown below.

Favorite School Lunch
20% pizza
27% chicken patty
15% cheeseburger
30% chicken nuggets
8% other

    a. What data display would best represent these results?

    b. Create a display of the data.

21. Place each function into one of the two categories.

Square Root	Cube Root

    $y = x^{1/3} + 7$

    $y = \sqrt[3]{x + 4} - 3$

    $y = -\sqrt{x - 3} + 8$

    $y = x^{1/2} - 4$

    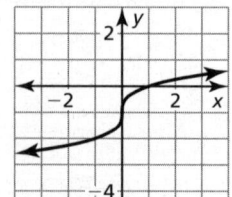

    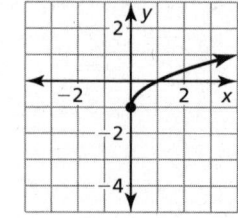

Chapters 9–11 Cumulative Test (continued)

22. A classmate created the following box-and-whisker plot using the data in the table.

Data
2, 4, 6, 2, 8, 12, 10, 14, 7

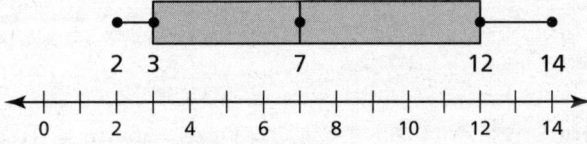

   a. Describe the error your friend made in creating the box-and-whisker plot of the data.

   b. Create an accurate box-and-whisker plot of the data.

23. You conduct a survey that asks 81 students about their favorite past time. The two-way table shows the results of the survey.

	Favorite Past Time	Not Favorite Past Time	Total
Reading	17	24	
Watching TV	21	19	
Total			

   a. Complete the table.

   b. What is the relative frequency of reading being the favorite past time?

   c. What is the relative frequency of watching TV being the favorite past time?

Post Course Test

Solve the equation. Check your solution.

1. $x + 14 = -5$
2. $3x - 2 = 4 - x$

Solve the equation.

3. $-3 + 4|8n + 10| = 53$
4. $-23 = -3(-4b - 3) - 8(1 + b)$

Describe the values of *c* for which the equation has no solution.

5. $-3x + c = -3x + 8$
6. $-|3 + x| = c$

Write the sentence as an inequality.

7. The quotient of *n* and 3 is less than 5.

8. 10 more than *y* is greater than or equal to 17.

Solve the inequality. Graph the solution.

9. $-3 \leq -3(1 - x)$
10. $-|3 + 2x| + 1 \leq -2$

Write and graph a compound inequality that represents the numbers that are *not* solutions of the inequality represented by the graph shown.

11.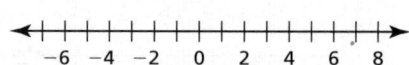

12.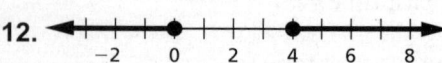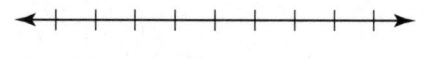

Determine whether the relation is a function. If the relation is a function, determine whether the function is *linear* or *nonlinear*.

13.
x	-3	0	3	6	9
y	-3	-4	-5	-6	-7

14. $y = -(x + 1)^2$

Write an equation in slope-intercept form of the line with the given characteristics.

15. through: $(-3, 1)$, slope $= 2$

16. through: $(-3, -4)$, perpendicular to $y = -\frac{3}{4}x + 5$

Answers

1. _____
2. _____
3. _____
4. _____
5. _____
6. _____
7. _____
8. _____
9. See left. _____
10. _____
See left. _____
11. _____
See left. _____
12. _____
See left. _____
13. _____
_____
14. _____
_____
15. _____
16. _____

Name_____ Date_____

Post Course Test (continued)

Write an equation in point-slope form of the line with the given characteristics.

Answers

17. through: $(-1, -3)$ and $(2, 5)$

17. _____

18. through: $(4, 4)$, parallel to $y = -\frac{1}{6}x + 2$

18. _____

19. ___See left.___

Graph the equation and identify the intercept(s). If the equation is linear, find the slope of the line.

19. $2x + 5y = 10$ 20. $y = 2|x - 1| - 4$

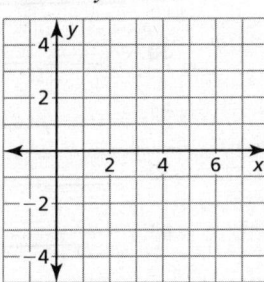 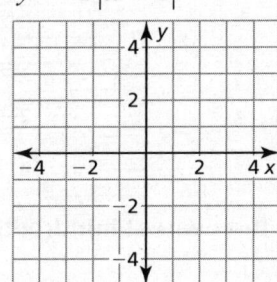

20. ___See left.___

21. ___See left.___

Graph f and g. Describe the transformations from the graph of f to the graph of g.

22. ___See left.___

21. $f(x) = x$; $g(x) = -3x + 2$ 22. $f(x) = |x|$; $g(x) = -|2x - 1| + 1$

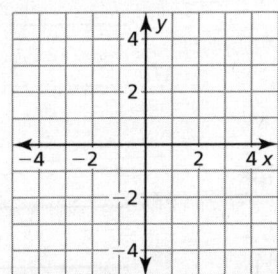 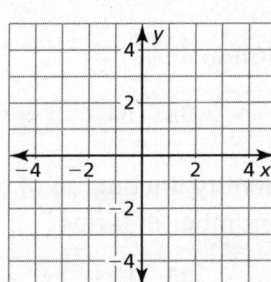

23. ___See left.___

Graph the function. Describe the domain and range.

24. ___See left.___

23. $f(x) = \begin{cases} -\frac{2}{3}x, & \text{if } x < 3 \\ 6, & \text{if } x \geq 3 \end{cases}$ 24. $y = \begin{cases} x, & \text{if } x \leq -3 \\ -3, & \text{if } -3 < x < 2 \\ 3x - 6, & \text{if } x \geq 2 \end{cases}$

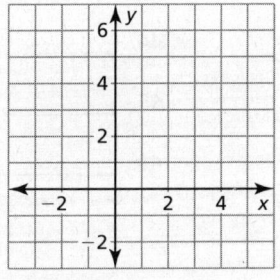 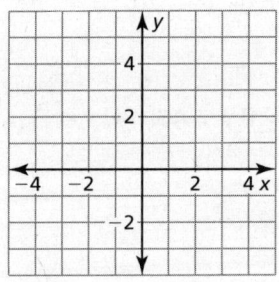

Name _____ Date _____

Post Course Test (continued)

Solve the system of linear equations using any method.

25. $2x + 6y = 14$
    $2x - y = -7$

26. $-x = -6y + 19$
    $4 = -4x - 12y$

Graph the system of linear inequalities.

27. $y \geq x + 1$
    $\frac{1}{3}x + y \geq -3$

28. $4x - y \leq -1$
    $y < -3$

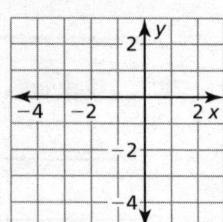
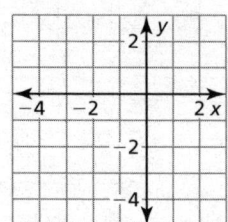

Evaluate the expression. Round to the nearest hundredth, if necessary.

29. $\sqrt[6]{100}$

30. $(36)^{1/4}$

31. $(-27)^{4/3}$

Simplify the expression. Write your answer using only positive exponents.

32. $x^3 \cdot x^{-7}$

33. $\dfrac{2w^{-3}z^{-2}}{4w^{-4}z^2}$

34. $-\left(-\dfrac{4a^2}{3ab^{-1}}\right)^{-3}$

Solve the equation. Check your solution.

35. $16^{-r-1} = \dfrac{1}{4}$

36. $64^{-2x+3} = 1$

Find the sum or difference. Then identify the degree of the sum or difference and classify the polynomial by the number of terms.

37. $(3x^2 + 6x) + (4x^2 - 8x)$

38. $(3 + 7x^3 + x^4) - (8 - x + x^4)$

Find the product.

39. $(c - 5)(c - 3)$

40. $(2a + 7)(7a - 4)$

41. $(2x + 1)(2x - 1)$

Factor the polynomial completely.

42. $b^3 - 3b^2 + b - 3$

43. $-n^2 + n + 20$

44. $2x^2 - 17x + 21$

Solve the equation.

45. $-2x(3x + 1)(x - 8)(x + 2) = 0$

46. $4k^2 + 3 = -13k$

Answers

25. _____
26. _____
27. See left.
28. See left.
29. _____
30. _____
31. _____
32. _____
33. _____
34. _____
35. _____
36. _____
    _____
37. _____
    _____
38. _____
39. _____
40. _____
41. _____
42. _____
43. _____
44. _____
45. _____
46. _____

Name_____ Date_____

Post Course Test (continued)

Solve the equation using any method.

47. $-3x^2 = -54$ 48. $x^2 - 8x + 7 = 0$ 49. $-3x^2 - 2x + 4 = 0$

Graph the function. Compare the graph to the graph of $f(x) = x^2$.

50. $g(x) = 3x^2 - 1$

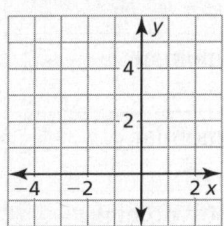

51. $h(x) = \frac{1}{2}(x + 4)^2$

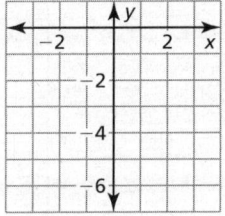

Find the inverse of the function.

52. $f(x) = -2 - \frac{1}{4}x$ 53. $f(x) = \sqrt{x + 3} - 4$

Graph the function f. Describe the domain and range. Compare the graph of f to the graph of g.

54. $f(x) = \sqrt{-x + 2}$; $g(x) = \sqrt{x}$ 55. $f(x) = \sqrt[3]{x} - 4$; $g(x) = \sqrt[3]{x}$

Solve the equation. Check your solutions.

56. $8\sqrt{n - 2} = -16$ 57. $\sqrt{2v} = \sqrt{3v - 1}$ 58. $2 + \sqrt{4x - 12} = x$

Find the mean, median, mode, and range of the data set. Round to the nearest tenth, if necessary.

59.

Shoe Sizes			
8.5	7.5	7	6
8.5	9.5	10	13
10	11.5	10	12

60. 20.1, 30.5, 22.3, 19.7, 17.5, 32.1

Answers

47. _____
48. _____
49. _____
50. _See left._
51. _See left._
52. _____
53. _____
54. _See left._
55. _See left._
56. _____
57. _____
58. _____
59. _____
60. _____

Post Course Test Item Analysis

Item Number	Skills
1	solving equations
2	solving equations
3	solving equations
4	solving equations
5	number sense
6	number sense
7	writing inequalities
8	writing inequalities
9	solving inequalities
10	solving inequalities
11	writing and graphing inequalities
12	writing and graphing inequalities
13	understanding functions
14	understanding functions
15	writing linear equations
16	writing linear equations
17	writing linear equations
18	writing linear equations
19	graphing linear equations
20	graphing linear equations
21	comparing linear transformations
22	comparing linear transformations
23	graphing piecewise functions
24	graphing piecewise functions

Item Number	Skills
25	solving linear systems
26	solving linear systems
27	graphing linear inequalities
28	graphing linear inequalities
29	evaluating rational expressions
30	evaluating rational expressions
31	evaluating rational expressions
32	simplify exponents
33	simplify exponents
34	simplify exponents
35	solving exponential equations
36	solving exponential equations
37	simplify polynomial expressions
38	simplify polynomial expressions
39	simplify polynomial expressions
40	simplifying polynomial expressions
41	simplify polynomial expressions
42	factoring
43	factoring
44	factoring
45	solving quadratics
46	solving quadratics
47	solving quadratics
48	solving quadratics
49	solving quadratics

Post Course Test Item Analysis (continued)

Item Number	Skills
50	graphing quadratics
51	graphing quadratics
52	finding inverses
53	finding inverses
54	graphing radicals
55	graphing radicals

Item Number	Skills
56	solving radical equations
57	solving radical equations
58	solving radical equations
59	measures of center
60	measures of center

Answers

Prerequisite Skills Test

1. -24 2. -1 3. -3 4. -17
5. 21 6. -2 7. -14 8. -16
9. 3 10. -5 11. 5 12. 7
13. 18 ft   14. 81 cm^2

15.
16.
17.
18.

19. $<$ 20. $<$ 21. $>$ 22. $>$
23. 12 24. 20 25. 73 26. -16
27. 83 28. -49

29.–32.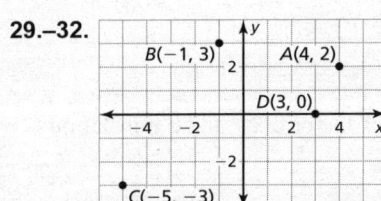

29. Quadrant I   30. Quadrant II
31. Quadrant III   32. x-axis
33. $(-2, 0)$ 34. $(-5, -2)$ 35. Q
36. $y = 2x - 3$ 37. $y = -\frac{3}{2}x - 2$
38. $y = -\frac{1}{3}x - \frac{1}{2}$ 39. $y = 7x + 12$
40. $y = \frac{1}{6}x + 1$
41. $p > 3$;
42. $x < 2$;
43. $m \geq -4$;

44. $x \leq -5$

45. 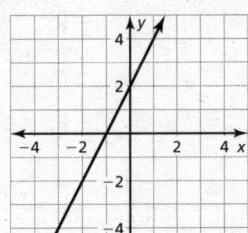   46.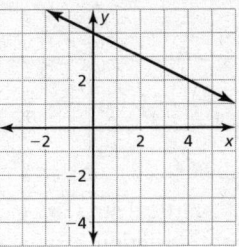

47.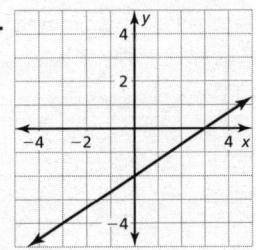

48. -9 49. 5 50. 5 51. -9
52. ± 3 53. -12 54. $a_n = 3n$
55. $a_n = -7n + 14$ 56. $a_n = 11n - 9$
57. $9x - 1$ 58. $-3m + 9$
59. $-5y - 3$ 60. $12d + 18$
61. 21 62. 37 63. 5 64. 8
65. -27 66. -30 67. $(-1, -1)$ 68. $(0, 1)$
69. $(3, -2)$ 70. 9 71. 16 72. 5
73. 1 74. $(x + 2)^2$ 75. $(x - 5)^2$
76. $(x + 4)^2$ 77. $(x - 13)^2$ 78. $(x + 3)^2$
79. $(x - 11)^2$ 80. 3 81. $-\frac{20}{3}$
82. 13 83. 3

84.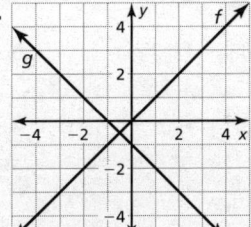

reflection in the y-axis, followed by a translation 1 unit down

Answers

85.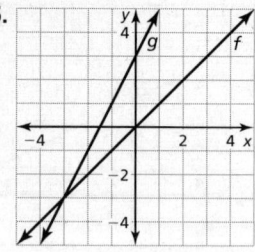

vertical stretch by a factor of 2, followed by a translation 3 units up

86.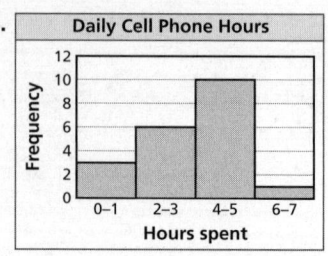

Pre-Course Test Answers

1. $x = 21$ **2.** $x = -12$ **3.** $p = 0, \frac{4}{9}$

4. $x = -1$

5. all real numbers except -4

6. $c < 0$ **7.** $m + 12 < 48$ **8.** $10x \geq 23$

9. $x \leq 5$
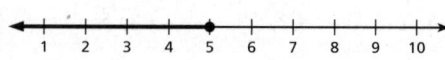

10. $-4 \leq x \leq 8$
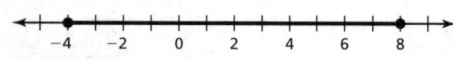

11. $x < -1$ or $x \geq 5$

12. $0 < x < 4$

13. yes; nonlinear   **14.** yes; linear

15. $y = -\frac{7}{6}x + \frac{1}{2}$ **16.** $y = 2x + 11$

17. $y + 3 = -\frac{1}{5}(x - 5)$ **18.** $y - 2 = -\frac{3}{2}(x + 4)$

19.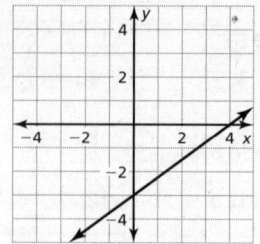

x-int: $(4, 0)$; y-int: $(0, -3)$; slope $= \frac{3}{4}$

20.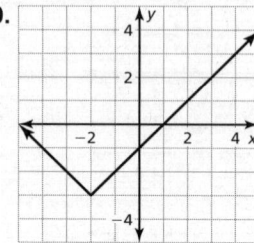

x-int: $(1, 0), (-5, 0)$; y-int: $(0, -1)$

21.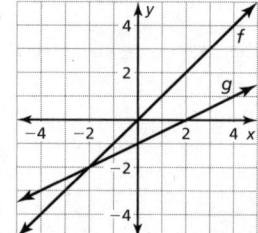

vertical shrink by a factor of $\frac{1}{2}$ and a translation 1 unit down

22.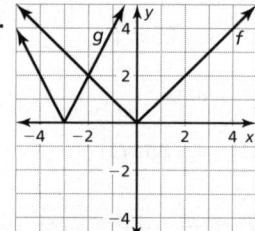

vertical stretch by a factor of 2 and a translation 3 units left

23.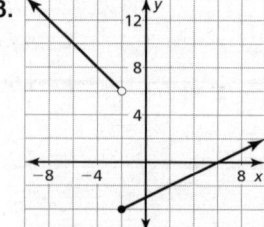

domain: all real numbers; range: $y \geq -4$

Answers

24.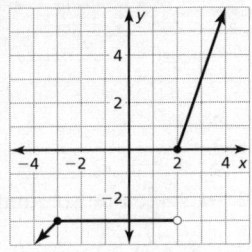
domain: all real numbers; range: $y \geq -2$

25. $(-9, 6)$ **26.** $(3, 4)$

27. 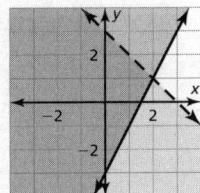 **28.**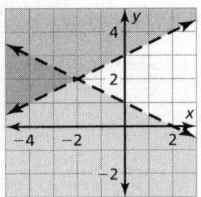

29. 2.65 **30.** -2.45 **31.** 243

32. m **33.** $\dfrac{a}{b}$ **34.** $\dfrac{4y^2}{9x^6}$

35. $b = \dfrac{3}{2}$ **36.** $p = \dfrac{5}{6}$

37. $12x^3 + x^2 + 6$; 3; trinomial

38. $-3x^4 + 12x^2$; 4; binomial

39. $m^2 + 2m - 24$ **40.** $12a^2 - 6a - 36$

41. $x^2 - 49$ **42.** $(x^2 - 2)(x - 2)$

43. $(n - 3)(n - 1)$ **44.** $-(2v + 5)(v + 9)$

45. $x = 0, 4, -\dfrac{1}{2}$ **46.** $g = 5, -2$

47. $x = 7, -7$ **48.** $x = 5 \pm \sqrt{23}$

49. $x = \dfrac{-3 \pm \sqrt{17}}{4}$

50.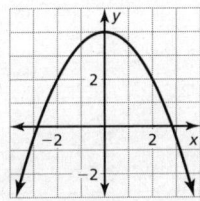
reflection in x-axis, a vertical shrink by a factor of $\dfrac{1}{2}$, and a translation 4 units up

51.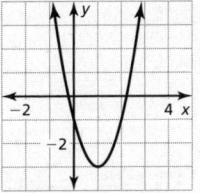
vertical stretch by a factor of 2, and a translation 1 unit right and 3 units down

52. $g(x) = -\dfrac{1}{2}x - \dfrac{5}{2}$ **53.** $g(x) = \sqrt{\dfrac{-x + 1}{2}} + 4$

54.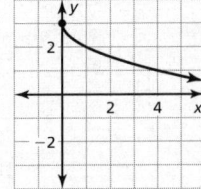
domain: $x \geq 0$; range: $y \leq 3$; reflected in x-axis and translated 3 units up

55.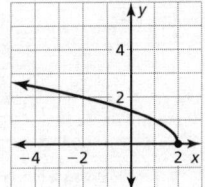
domain: all real numbers; range: all real numbers; translated 2 units left

56. $m = 24$ **57.** $n = 1$ **58.** $p = 4$

59. mean $= 82.7$, median $= 85.5$, mode $= 88$, range $= 34$

60. mean $= 4.8$, median $= 4$, mode $= 4$, range $= 10$

Chapter 1

1.1–1.3 Quiz

1. $x = -4$ **2.** $z = 8$ **3.** $r = -12$

4. $p = 75$ **5.** $m = 5$ **6.** $v = 3$

7. $w = 3$ **8.** $x = -16$

9. $c = -3$; one solution

10. infinitely many solutions

11. no solution

12. $y = 4$; one solution

Answers

13. $y \div 12 = 4560$; $54,720

14. $24h + 90 = 810$; 30 h

15. a. you, 12 mi   b. 6 h

Test A

1. $x = \frac{1}{4}$; Subtract $\frac{1}{2}$ from each side.

2. $z = 48$; Multiply each side by 4.

3. $n = -4$; one solution

4. $g = 12$; one solution

5. infinitely many solutions

6. $y = 8$; one solution

7. no solution

8. $x = 3$; one solution

9. infinitely many solutions

10. no solution

11. all real numbers except 6

12. $c < 0$ 13. $x = 15$; 75°, 15°, 90°

14. $b = 20$; 100°, 80°, 100°, 80°

15. $n = 4$ 16. $x = 6$ 17. $d = -42$

18. $b = 27, b = -3$ 19. $r = 5$

20. $k = -2, k = -6$ 21. $y = 2 - 5x$

22. $y = 4 - x$

23. a. $h = \dfrac{V}{\pi r^2}$ b. 2 in.

24. 20°   25. 4 books   26. $|x - 14| = 4$

27. a. 50 brochures
    b. Company B

Test B

1. $x = \frac{1}{6}$; Subtract $\frac{2}{3}$ from each side.

2. $w = 20$; Add 8 to each side.

3. $m = -12$; one solution

4. $n = 45$; one solution

5. no solution   6. $h = \frac{9}{2}$; one solution

7. infinitely many solutions

8. no solution   9. $d = -\frac{3}{2}$; one solution

10. $w = 8$; one solution

11. $c = 4$ 12. $c = 6$

13. $x = 40$; 85°, 120°, 90°, 65°

14. $x = 20$; 38°, 40°, 102°

15. $n = -3$ 16. $x = -7$

17. $w = 6$ 18. $m = 4, m = -20$

19. $y = 1$ 20. $k = -7, k = -\frac{3}{7}$

21. $y = 6 - \frac{3}{2}x$ 22. $y = x + 2$

23. a. $h = \dfrac{3V}{\pi r^2}$

    b. $h = 10$ cm

24. width = 10 m; length = 15 m

25. 5 beads

26. The right side would be negative, but the absolute value on the left side indicates a value that is not negative.

27. a. 15 T-shirts   b. Company A

Alternative Assessment

1. a. one solution; $x = 15$
   b. infinitely many solutions
   c. two solutions; $x = 2, -\frac{10}{3}$
   d. no solution
   e. no solution
   f. two solutions; $x = 2, 0$

Answers

2. a. $h = \dfrac{2A}{b_1 + b_2}$

b. $2A = 2\left(\dfrac{1}{2}h\right)(b_1 + b_2)$

$2A = h(b_1 + b_2)$

$\dfrac{2A}{h} = b_1 + b_2$

$b_1 = \dfrac{2A}{h} - b_2$

c. $2A = 2\left(\dfrac{1}{2}h\right)(b_1 + b_2)$

$2A = h(b_1 + b_2)$

$2A = b_1h + b_2h$

$2A - b_2h = b_1h$

$b_1 = \dfrac{2A - b_2h}{h}$

d. $b_1 = \dfrac{2A - b_2h}{h}$

$b_1 = \dfrac{2A}{h} - \dfrac{b_2h}{h}$

$b_1 = \dfrac{2A}{h} - b_2$

e. You can solve for b_2 using the same methods that you used to solve for b_1.

Chapter 2

2.1–2.4 Quiz

1. $q + 8 \leq 15$ **2.** $20 \geq 4c - 8$

3. $x > -4$ **4.** $x \leq 9$

5. $t \leq 7$ **6.** $m < 4$

7. $s \geq 4$ **8.** $p > -27$

9. $y \geq 6$ **10.** $t > -6$

11. all real numbers **12.** no solution

13. a. $a \geq 18, w > 1$

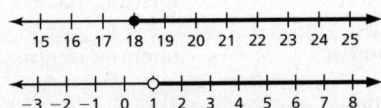

b. no

14. $s + 400 < 1500;\ s < 1100$

Test A

1. $2n + 8 \leq 25$ **2.** $t \geq 75$

3. $t \leq 26$ **4.** $x > 2$ **5.** $x \leq 3$

6. $m > -3$

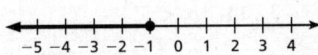

7. $z \leq -1$

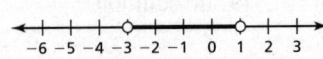

8. $m \leq 1$ **9.** $x \geq -\dfrac{8}{3}$

10. no solution **11.** $k > 5$

12. all real numbers **13.** $x < \dfrac{1}{6}$

14. $y < -3$ or $y > 4$

15. $-3 < c < 1$

16. $a < 5$ or $a > 6$ **17.** $-4 < g < 1$

18. no solution **19.** $y \leq 1$ or $y \geq 6$

20. $x < -2$ or $x \geq 3$

21. $-1 \leq x \leq 1$

22. $w + 285 \geq 500;\ w \geq 215$

23. $2.5b \geq 30;\ b \geq 12$

24. $4t - 180 \geq 500;\ t \geq 170$

25. $30 + 7.50h \leq 115;\ h \leq 11\dfrac{1}{3}$

Test B

1. $2n \geq 14$ **2.** $s \leq 60$

3. $r \geq 28$ **4.** $x > 5$ **5.** $-1 < x < 5$

Answers

6. $x \leq -7$

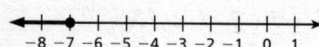

7. $q > -7$

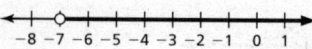

8. no solution   **9.** $p > -6$

10. $w < 20$ **11.** $p < 2$

12. all real numbers   **13.** all real numbers

14. $y < 1.5$ or $y > 3.5$

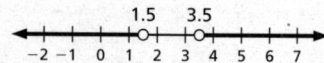

15. $-5 < c < 1$

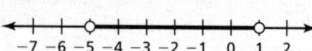

16. $p \geq 4$ or $p < -24$ **17.** no solution

18. $-7 < x < -3$ **19.** no solution

20. $x < 0$ or $x \geq 5$

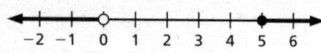

21. $1 \leq x \leq 3$

22. $6h \geq 75;\ h \geq 12.5$

23. $150 \leq 25b \leq 225;\ 6 \leq b \leq 9$

24. $\dfrac{245 + m}{7} \geq 50;\ m \geq 105$

25. $1500 + 250h \leq 5000;\ h \leq 14$

Alternative Assessment

1. c, d, and f; b, e, and g; a

2. a. $40 < 3x - 5 \leq 130$
   b. $15 < x \leq 45$
   c.
   d. $55 < 2x - 15 \leq 105$
   e. $35 < x \leq 60$
   f.

g. yes; The graphs overlap, so there is a common solution set.

h. $35 < x \leq 45$

Chapter 3

3.1–3.3 Quiz

1. function; Every input has exactly one output.

2. not a function; The input 3 has two outputs.

3. domain: $-1, 0, 1, 4$; range: $-2, -1, 0, 2$

4. domain: all real numbers; range: $y \geq -2$

5. domain: $-2 \leq x \leq 2$; range: $-2 \leq y \leq 3$

6. nonlinear; The graph is not a line.

7. linear; The rate of change is constant.

8. nonlinear; When distributing the x, you obtain $y = x^2 + 3x$. Because the degree of the equation is greater than 1, it is nonlinear.

9.   **10.**

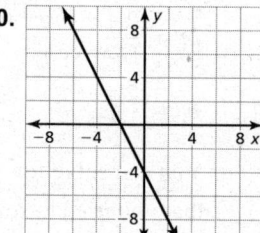

11.

12. independent variable: h, dependent variable: c; continuous; It is possible to use the lawn care service for a fraction of an hour.

Test A

1. not a function   **2.** function; linear

3. function; linear   **4.** function; nonlinear

5. domain: all real numbers; range: $y \geq 0$; continuous

6. domain: $-2, -1, 0, 3$; range: $-2, -1, 0, 3$; discrete

Answers

7. $g(-1) = 4, g(0) = 1, g(4) = 49$

8. $b(-1) = -2, b(0) = -4, b(4) = -12$

9. $h(-1) = 6, h(0) = 5, h(4) = 1$

10. 2   11. 1

12. x-intercept: $(3, 0)$, y-intercept: $(0, 2)$

13. x-intercept: $(10, 0)$, y-intercept: $(0, -6)$

14. x-intercept: $(-16, 0)$, y-intercept: $(0, -8)$

15. $m = -\frac{3}{2}$

16. $m =$ undefined

17. 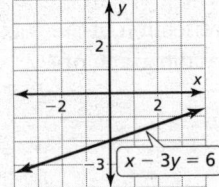   18.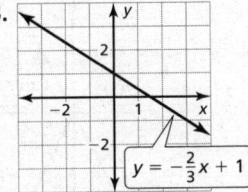

19. $m = -\frac{5}{3}$, y-intercept: $(0, 5)$, x-intercept: $(3, 0)$

20. $m = 1$, y-intercept: $(0, -3)$, x-intercept: $(3, 0)$

21. m is undefined, y-intercept: none,
    x-intercept: $(-4, 0)$

22. *Sample answer:* The graph of g is a horizontal translation 3 units left of the graph of f.

23. The graph of g is a vertical stretch of the graph of f by a factor of 2.

24. **a.** The graph of g is a reflection in the x-axis, followed by a translation 2 units right and 3 units up of the graph of f.

    b.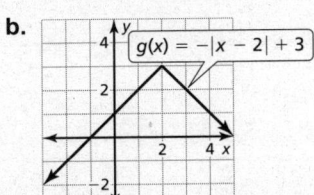

Test B

1. function; nonlinear   2. function; linear

3. function; linear   4. function; nonlinear

5. domain: all real numbers, range: $y \geq 0$; continuous

6. domain: $-1, 0, 2, 4$; range: $-2, -1, 3$; discrete

7. $g(-3) = -16, g(-2) = -11, g(1) = -8$

8. $h(-3) = 0, h(-2) = 2, h(1) = 8$

9. $f(-3) = -\frac{5}{2}, f(-2) = -2, f(1) = -\frac{1}{2}$

10. 8   11. $\frac{9}{4}$ 12. 9   13. -2

14. x-intercept: $(-5, 0)$, y-intercept: $\left(0, \frac{10}{3}\right)$

15. x-intercept: $(-4, 0)$, y-intercept: $\left(0, -\frac{8}{5}\right)$

16. x-intercept: $(-18, 0)$, y-intercept: $(0, 6)$

17. 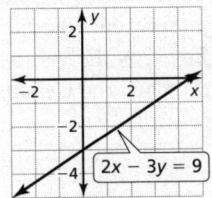   18.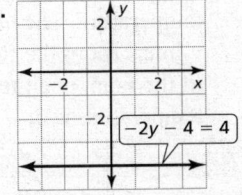

19. $m = 0$ 20. $m = \frac{3}{2}$

21. **a.** independent: m miles; dependent: c cost
    b. $130
    c. 165 mi

22. $m = -1$, y-intercept: $(0, 3)$, x-intercept: $(3, 0)$

23. $m = \frac{2}{3}$, y-intercept: $\left(0, -\frac{7}{3}\right)$, x-intercept: $\left(\frac{7}{2}, 0\right)$

24. $m = 0$, y-intercept: $\left(0, -\frac{14}{3}\right)$,
    x-intercept: none

25. The graph of g is a reflection in the y-axis of the graph of f.

26. The graph of g is a vertical shrink of the graph of f by a factor of $\frac{1}{2}$.

Answers

27. a. The graph of g is a reflection in the x-axis, followed by a vertical stretch by a factor of 2 and a translation 1 unit right and 2 units up of the graph of f.

b.
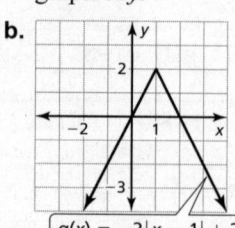
$g(x) = -2|x - 1| + 2$

Alternative Assessment

1. a. independent variable: C, dependent variable: F

b. $[-273.15, \infty)$; continuous; The temperature can be any real number in the interval.

c. The number $\frac{9}{5}$ represents the slope of the function. The Fahrenheit temperature raises 9 degrees for every 5-degree increase in Celsius temperature. The number 32 represents the y-intercept of the function. So, 32 is the value of the Fahrenheit temperature when the Celsius temperature is 0.

d.
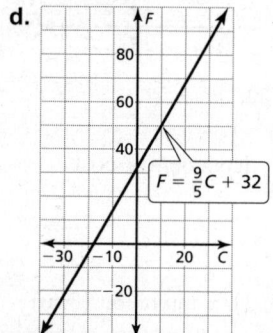
$F = \frac{9}{5}C + 32$

e. $C = 30°, F = 86°; C = -15°, F = 5°;$
$C = -22°, F = -7.6°$

f. $21.67°$

g. Sample answer: $C = -10°, F = 14°$

h. $F(C) = \frac{9}{5}C + 32$

2. a.
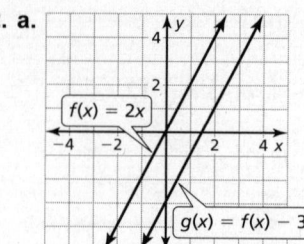
$f(x) = 2x$; $g(x) = f(x) - 3$

The graph of g is a horizontal translation 3 units down of the graph of f.

b.
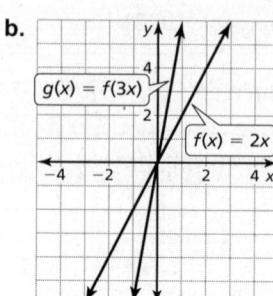
$g(x) = f(3x)$; $f(x) = 2x$

The graph of g is a horizontal shrink by a factor of $\frac{1}{3}$ of the graph of f.

c.
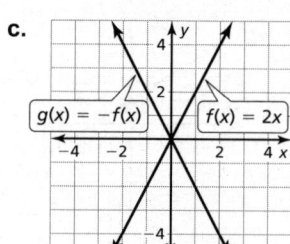
$g(x) = -f(x)$; $f(x) = 2x$

The graph of g is a reflection in the x-axis of the graph of f.

d.
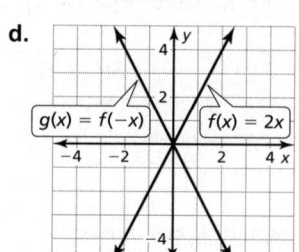
$g(x) = f(-x)$; $f(x) = 2x$

The graph of g is a reflection in the y-axis of the graph of f.

e.
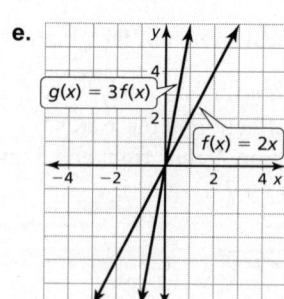
$g(x) = 3f(x)$; $f(x) = 2x$

The graph of g is a vertical stretch by a factor of 3 of the graph of f.

Answers

f. 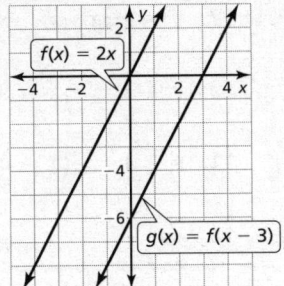 The graph of g is a horizontal translation 3 units right of the graph of f.

g.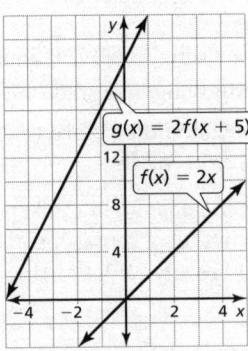

The graph of g is a vertical stretch by a factor of 2, followed by a horizontal translation 5 units left of the graph of f.

h.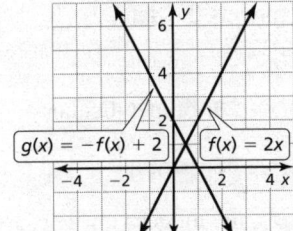

The graph of g is a reflection in the x-axis, followed by a vertical translation 2 units up of the graph of f.

i.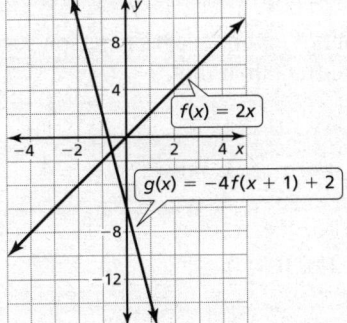

The graph of g is a vertical stretch by a factor of 4, followed by a reflection in the x-axis, and a translation 1 unit left and 2 units up of the graph of f.

Cumulative Test

1. C   2. B   3. C   4. A

5. B, C, D   6. D   7. B   8. D

9. C   10. A   11. C   12. C

13. D   14. A   15. C   16. D

17. B   18. A   19. C   20. D

21. A   22. C   23. D   24. A

25. $-3|1 - y| = -9$; The other three have no solution.

26. $15 - 2x = 3x$, $x + 2 = 5$, $4(x - 2) = 3x - 5$

27. a. $h = \dfrac{V}{\pi r^2}$

     b. $h = 5$ cm

28. a. $x + 32 \leq 50$; $x \leq 18$

     b. yes; The books add to 11 pounds, and you can add up to 18 more pounds.

29.

No solution	One solution
$3(-2 - 3x) = -9x - 4$	$1 - x = 6 - 6x$
$\|2x - 5\| + 3 = 0$	$\frac{1}{2}(6 - x) = x$

Two solutions	Infinitely many solutions
$\|x - 5\| = 4$	$8x + 12 = 4(2x + 3)$
$\|3x + 6\| = 9$	$3(x + 1) - 4 = 3x - 1$

30. a. 90%

     b. no; You would need to score 125% on the last quiz.

31. a. $36 \leq x \leq 54$

     b. $x \geq 4\frac{1}{2}$ ft

32. a. $|t - (-28.9)| \leq 2.8$; $-31.7 \leq t \leq -26.1$

     b. You would save money on the electric bill; The thermostat could be off, and the ice cream could be ruined.

Answers

33. a. $f(m) = 100m$

   b. domain: $x \geq 0$; range: $y \geq 0$

   c. continuous; You could run part of a mile, and you would still burn calories.

34. $y - b = mx$; $b = y - mx$; $x = \dfrac{y-b}{m}$; $m = \dfrac{y-b}{x}$

Chapter 4

4.1–4.3 Quiz

1. $y = -\frac{4}{3}x - 2$
2. $y = \frac{3}{2}x + 5$
3. $y = -\frac{2}{3}x + 2$
4. $y - 1 = -1(x - 4)$
5. $y - 3 = \frac{3}{4}(x - 1)$
6. $y + 5 = \frac{2}{3}(x + 2)$
7. $f(x) = -x + 2$
8. $f(x) = \frac{1}{3}x - 3\frac{2}{3}$
9. $a \parallel b$; The slopes of lines a, b, and c are 3, 3, and $-\frac{1}{10}$, respectively.
10. $a \parallel c$, $a \perp b$, $c \perp b$; The slopes of lines a, b, and c are 3, $-\frac{1}{3}$, and 3, respectively.
11. a. $c(m) = 800m + 1000$

    b. $10,600

    c. the first building; You have enough to rent the first building for 17.5 months, or the second building for 15 months.
12. yes; The rate of change is constant; $y = 5x$

Test A

1. domain: all real numbers, range: all real numbers
2. domain: $x < 3$, range: $y = -2$ or $3 \leq y \leq 5$
3.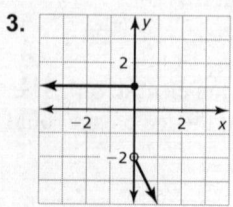
4. $y = \frac{1}{4}x + 2$
5. $y = -\frac{3}{2}x + 1$

6. $y = -\frac{3}{2}x - 2$
7. $y = -3x - 7$
8. $y = -2x + 8$
9. $y - 3 = 2(x - 0)$
10. $y - 5 = -2(x + 3)$
11. $y + 3 = \frac{3}{5}(x - 0)$
12. $y - 10 = \frac{1}{2}(x + 3)$
13. yes; -7
14. no
15. yes; -4
16. no
17. a. $r = 0.96$; There is a strong linear relationship between time spent studying and grades.

    b. Answers will vary but should be close to the line of best fit, $y = 6.315x + 62.045$.

    c. The slope represents the change in grade per hour studying; the y-intercept represents the grade a student would receive if he or she studied for 0 hours.
18. a. $y = 6.315x + 62.045$

    b. The correlation coefficient is 0.956. The relationship between time spent studying and grade has a strong positive correlation, and the equation closely models the data.

    c. 81
19. neither
20. perpendicular
21. neither
22. parallel
23. yes; The taller you are, the longer your legs are, so your strides will be longer.
24. no; There is no relationship between the number of flat tires on your car and the number of pets you own.
25. no; The amount of text messages a person sends daily does not affect their diet.

Test B

1. domain: $x < 4$ or $x \geq 5$, range: $y > -4$
2. domain: $x < 1$ or $1 < x < 4$, range: $y = -2$ or $0 < y \leq 3$
3.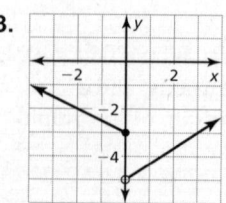

Answers

4. $y = \frac{2}{5}x + \frac{11}{5}$ **5.** $y = 5$

6. $y = 2x + 7$ **7.** $y = \frac{2}{3}x - 2$

8. $y = 5$ **9.** $y = \frac{1}{2}(x - 3)$

10. $y + 7 = -3(x - 4)$ **11.** $y - 6 = \frac{2}{5}(x - 7)$

12. $y - 1 = -\frac{1}{3}(x + 4)$ **13.** no

14. yes; -6 **15.** yes; $\frac{1}{3}$ **16.** no

17. line ℓ: $y = \frac{2}{5}(x + 3) - 3$ or $y = \frac{2}{5}x - \frac{9}{5}$

18. a.

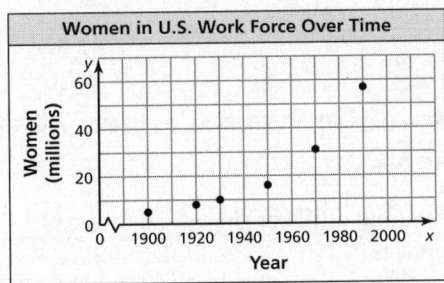

The graph suggests a positive correlation.

b. $y = 0.552x - 1052.3$

c. The correlation coefficient is 0.927, suggesting that there is a strong positive correlation between the number of women in the work force and the year. However, the scatter plot does appear to curve upward at the end.

19. neither **20.** parallel

21. perpendicular **22.** perpendicular

23. yes; The number of gallons of gas in the tank decreases as more miles are driven.

24. no; The height of a person has no relationship to the length of their hair.

Alternative Assessment

1. a. $m = \frac{1}{3}$

b. Sample answer: $y - 1 = \frac{1}{3}(x - 1)$

c. $y = \frac{1}{3}x + \frac{2}{3}$

d. $y = \frac{1}{3}x - 2$

e. $y = -3x + 1$

2. a. $y = 2.04x + 12.63$

b. The correlation coefficient is about 0.9135. This is close to ± 1, which means that the relationship between the height and the yield of strawberry plants has a strong positive correlation, and the equation closely model the data.

c. The points are evenly dispersed about the horizontal axis. So, the equation is a good fit.

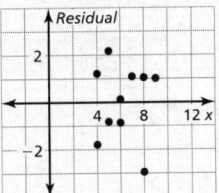

3. a.

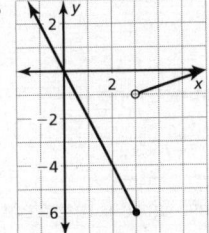

b. domain: all real numbers, range: $y \geq -6$

c. $y = -6$

d. $y = 0$

e. $y = -\frac{1}{3}$

Chapter 5

5.1–5.4 Quiz

1. $(1, 3)$ **2.** $(-4, 1)$ **3.** $(0, -2)$ **4.** $(2, 3)$

5. $(-3, -6)$ **6.** $(3, -4)$ **7.** $(6, -6)$ **8.** $(-1, 3)$

9. $(1, -2)$ **10.** $(4, -1)$ **11.** $(-3, 2)$ **12.** $(0, -3)$

13. a. $24x + 32y = 264$, $x + y = 9$

b. $(3, 6)$; You bought three shirts and six pairs of pants.

14. $0.06x + 0.15y = 69.84$, $x + y = 852$, $(644, 208)$

Test A

1. $(-1, -1)$ **2.** $(4, -3)$ **3.** $(3, -4)$ **4.** $(0, -3)$

5. $(-2, -4)$ **6.** $(-3, 2)$

Answers

7. 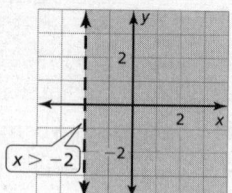   **8.**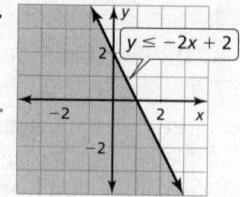

9. 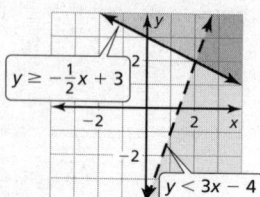   **10.**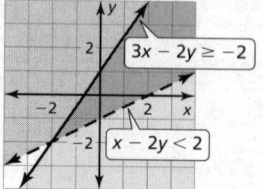

11. yes; pencils: $1.50, pens: $1.75

12. no solution; same slope $(m = 1)$, different y-intercepts

13. infinitely many solutions; same slope $\left(m = -\frac{2}{3}\right)$ and same y-intercept $(b = -2)$

14. one solution; different slopes, same y-intercept $(b = 7)$

15. a. $y \geq 5,\ 4x + 10y \leq 100$

b.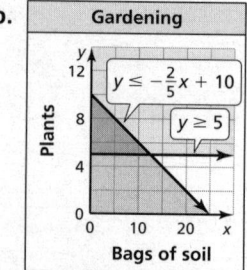

c. *Sample answer:* (10, 6); You can buy 10 bags of soil and 6 plants.

16. $y > -1,\ x \leq 3$

17. $y \geq \frac{2}{3}x + 3,\ y > -\frac{4}{3}x - 3$

18. $x = 1$ **19.** $x = -2$ **20.** $x = 1, 2$

Test B

1. $(-5, 5)$ **2.** $(-9, 3)$

3. infinitely many solutions

4. no solution

5. (0, 6)   **6.** (9, −10)

7. 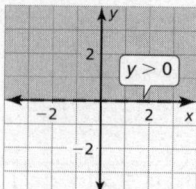   **8.**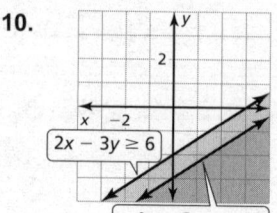

9. 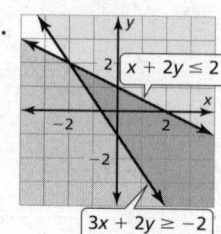   **10.**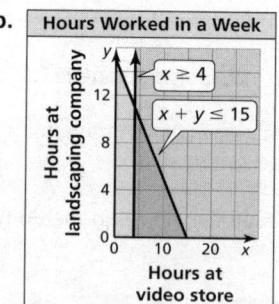

11. $y + 1$

12. $0.01x + 0.05y = 8.80,\ y = 2x$; 80 pennies and 160 nickels

13. one solution; different slopes and different y-intercepts

14. infinitely many solutions; same slope $\left(m = -\frac{2}{3}\right)$ and same y-intercept $\left(b = \frac{11}{3}\right)$

15. no solution; same slope $\left(m = -\frac{1}{2}\right)$ and different y-intercept

16. a. $x \geq 4,\ x + y \leq 15$

b.

c. $P(x, y) = 5x + 7y$

d. (4, 9); You could work 4 hours at the video store and 9 hours at the landscaping company.

17. $y < x + 3,\ y < -2x + 6$

18. $y \leq \frac{1}{3}x + 1,\ y > \frac{1}{3}x - 2$

Answers

19. $x = 5$ **20.** $x = 2, 4$ **21.** $x = 1, 3$

Alternative Assessment

1. a. $\left(-\frac{5}{3}, 0\right)$

 b. $\left(-\frac{5}{3}, 0\right)$

 c. $\left(-\frac{5}{3}, 0\right)$

 d. *Sample answer:* elimination; You can multiply the second equation by 2 and add the equations to solve for y.

 e. $6x - 2y = -10$

2. a. $x = 2$
 b. $x = 8$
 c. $x = 1$ and $x = 2.2$
 d. $x = \frac{1}{3}$

3. a.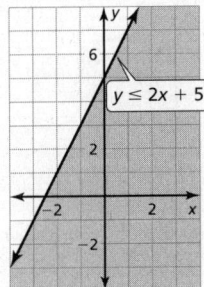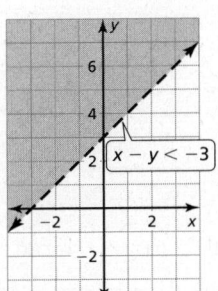

 b.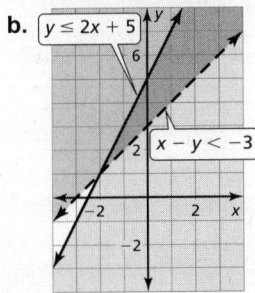

 c. no
 d. no
 e. yes

Cumulative Test

1. A **2.** $y = -2x + 3$

3. B

4. $y = \frac{1}{2}x$, $x - 2y = 0$, $y - 2 = \frac{1}{2}(x - 4)$

5. $y = -\frac{1}{3}x - 6$, $y = 3x - 26$

6. B **7.** D **8.** C **9.** D

10. A

11. a. $y = -\frac{25}{2}x + 100$

 b. 8 weeks

12. C **13.** B **14.** A

15. yes; $d = x$; $4x + 2, 5x + 2, 6x + 2$

16. a. $f(x) = \begin{cases} 2, & \text{if } x \leq -2 \\ x + 1, & \text{if } -2 < x < 3 \\ 4, & \text{if } x \geq 3 \end{cases}$

 b. 1
 c. 4

17. a. $y \geq 0$, $y \leq x + 4$, $y \leq -x + 4$
 b. 16 square units

18. a. $f(x) = \begin{cases} 0.49, & \text{if } 0 < x \leq 1 \\ 0.70, & \text{if } 1 < x \leq 2 \\ 0.91, & \text{if } 2 < x \leq 3 \\ 1.12, & \text{if } 3 < x \leq 3.5 \end{cases}$

 b. step function; The cost of the stamp is constant within each weight class.

 c.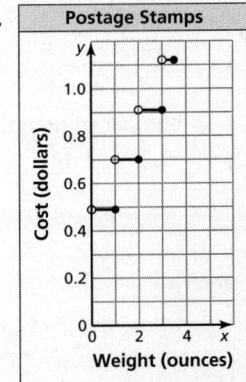

19. a. $p(t) = 1.25t + 10$
 b. $10
 c. The meat lover's pizza is $0.50 less.

Answers

20. a.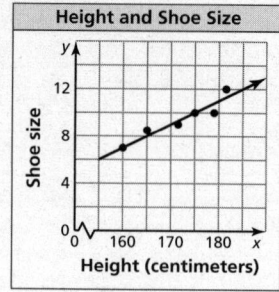

b. $y = 0.2x - 24$

c. The slope of 0.2 means shoe size increases by 0.2 for every 1 centimeter increase in height. The y-intercept of −24 has no meaning in this context because shoe size cannot be negative.

d. positive

e. size 19

f. The equation predicted a smaller shoe size for the player. It was inaccurate by three sizes.

g. $y = 0.2x - 24$

h. $r = 0.96$; strong positive correlation

21. a. $y = 500x + 250$

b. $2750

22. $y = -\frac{3}{2}x - 4$, $y - 1 = -\frac{3}{2}(x + 2)$, $3x = 9 - 2y$, $x = -\frac{2}{3}y + 2$

23. a. $x + y = 90$, $20x + 10y = 1300$

b. 40 min

24. a. x = number of pens, y = number of pencils, $x \geq 50$, $y \geq 100$, $1.25x + 0.25y \geq 115$

b.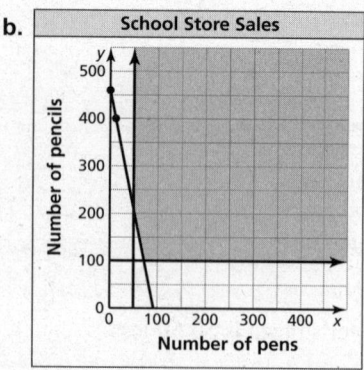

c. *Sample answer:* (70, 115), (80, 200)

d. no; doesn't satisfy the third inequality

25. one solution: $-8x + 10y = 24$
$6x + 5y = 2$;
$-7x + 2y = 18$
$6x + 6y = 0$;
$5x + 4y = -14$
$3x + 6y = 6$

no solution: $-3x + 3y = 4$
$-x + y = 3$;
$3 + 2x - y = 0$
$2y + 12 = 4x$

infinitely many solutions: $2x + 8y = 6$
$-5x - 20y = -15$

Chapter 6

6.1–6.4 Quiz

1. 4^5 2. $\dfrac{1}{k^6}$

3. $\dfrac{2}{r^2}$ 4. $\dfrac{x^6}{4y^4}$

5. 4   6. $\frac{1}{3}$

7. 1000   8. 27

9.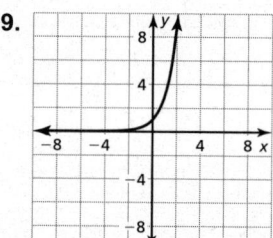

domain: all real numbers; range: $y > 0$

10. neither; terms do not change by a constant ratio

11. exponential decay; terms change by the constant ratio $\frac{1}{13}$

12. exponential growth; 65%

13. exponential growth; 20%

14. exponential decay; 60%

15. $12ab^3$ ft

Answers

16. a. exponential growth

b.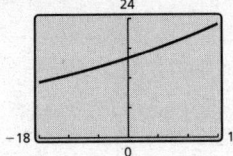

domain: all real numbers; range: $y > 0$

c. 40%   **d.** 3.3%   **e.** 120 deer

Test A

1. $\dfrac{y^4}{8}$ **2.** 1   **3.** $\dfrac{1}{125}$

4. $-64x^6$ **5.** $81a^{12}$ **6.** $\dfrac{12d^6}{c^7}$

7. -6 **8.** 1024   **9.** 4

10.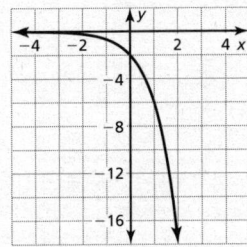

domain: all real numbers; range: $y < 0$

11.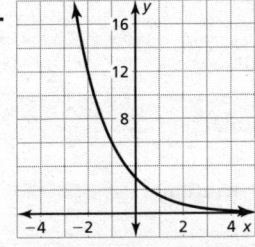

domain: all real numbers; range: $y > 0$

12. $x = -4$ **13.** $x = 9$

14. a. $y = 500(1.07)^t$

b. $572.45

15. a. $y = 6599(0.88)^t$

b. $4793.85

c. $511.84

d. never; An exponential function will approach zero, but can never equal zero.

16. neither   **17.** exponential decay

18. arithmetic   **19.** geometric

20. neither   **21.** geometric

22. $a_1 = 25;\ a_n = a_{n-1} - 15$

23. $a_1 = -10;\ a_n = a_{n-1} + 4$

24. -7 **25.** $y = 32$ **26.** $f(-1) = \dfrac{3}{4}$

27. $f(3) = \dfrac{125}{2}$ **28.** $y = 16$

29. a.

Minutes, x	0	1	2	3	4	5	6
Number of bacteria, y	1	2	4	8	16	32	64

b. $y = 2^x$

c. 1,048,576

Test B

1. $\dfrac{108y^7}{x^3}$ **2.** $\dfrac{1}{125x^{12}}$ **3.** $-\dfrac{8}{a^6}$

4. Sample answer: $(3x^3y^4)^2$

5. Sample answer: $(4x^3y^3)^3$

6. $\dfrac{1}{9}$ **7.** 5   **8.** $\dfrac{4}{3}$

9.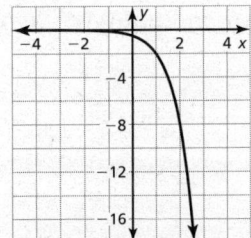

domain: all real numbers; range: $y < 0$

10.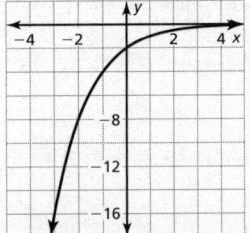

domain: all real numbers; range: $y < 0$

Answers

11. a. $y = 675(1.005)^{12t}$
  b. $807.76

12. a. $y = 12(2.2)^t$ in., or $y = 2.2^t$ ft
  b. 618.4 in., or 51.536 ft
  c. 7 times

13. $x = -4$ **14.** $x = 31$ **15.** neither

16. exponential decay

17. neither **18.** geometric **19.** arithmetic

20. geometric

21. $a_1 = 27; a_n = \frac{1}{3} a_{n-1}$

22. $a_1 = -2; a_n = 2a_{n-1} + 1$

23. $-\frac{17}{2}$ **24.** $\frac{1}{48}$ **25.** -1

26. a. exponential growth **b.** 50% growth
  c. 4.2% growth **d.** 6750 people

Alternative Assessment

1. a. $\dfrac{y^5}{x^6}$ **b.** $\dfrac{1}{x^{21}}$ **c.** $\dfrac{49x^{10}}{16y^6}$ **d.** 4 **e.** 9

2. a. $x = -1$ **b.** $x = 3$ **c.** $x = -2$

3. a. explicitly; geometric; 10, 20, 40, 80, 160, 320

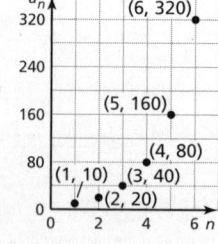

  b. explicitly; arithmetic; $-6, -1, 4, 9, 14, 19$

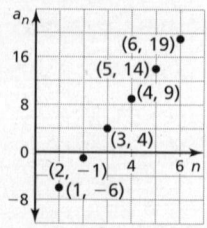

  c. explicitly; geometric; 2, $\frac{2}{3}, \frac{2}{9}, \frac{2}{27}, \frac{2}{81}, \frac{2}{243}$

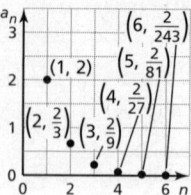

  d. recursively; arithmetic; 3, 5, 7, 9, 11, 13

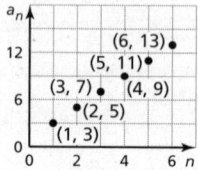

  e. recursively; geometric; 24, -12, 6, -3, $\frac{3}{2}$, $-\frac{3}{4}$

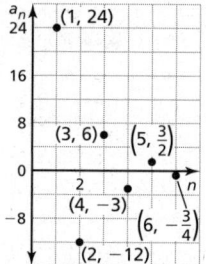

Chapter 7

7.1–7.4 Quiz

1. $-6r^4$; 4; -6; monomial

2. $g^2 - 2g + 4$; 2; 1; trinomial

3. $-\frac{3}{8}n^5 + \frac{1}{4}n^3$; 5; $-\frac{3}{8}$; binomial

4. $2a^4 + 8.1a^3 - 1.6a$; 4; 2; trinomial

5. $2x^2 + 9$ **6.** $-5n^2 + 2n + 5$

7. $-4h^2 + 7h - 12$ **8.** $-m^2 - mn + n^2$

9. $x^2 + 9x + 20$ **10.** $-6d^2 + 26d - 28$

11. $y^3 + 9y^2 + 14y - 24$

12. $25z^2 - 9$ **13.** $x = 0, 4$ **14.** $y = 6$

15. $p = -\frac{3}{4}, \frac{5}{2}, -2$ **16.** $y = 0, 9, -\frac{2}{3}$

17. a. $8x + 72$
  b. $4x^2 + 72x + 320$
  c. 88 in., 480 in.^{2}

Answers

Test A

1. $x^2 - 7x + 5$; 2; 1; trinomial
2. $-7v^4$; 4; -7; monomial
3. $-1.5z^7 + 5z^6 + z$; 7; -1.5; trinomial
4. $\frac{2}{3}a^5 - \frac{1}{2}a^4$; 5; $\frac{2}{3}$; binomial
5. $2b^3 - 3b^2 - 12b$
6. $-2x^2 + 5$
7. $-11x^2 + 11x - 5$
8. $-6y^3 - 2y^2 + 7y + 4$
9. $a^2 + 2a - 15$
10. $6c^2 + 5c - 4$
11. $x^3 + 13x^2 + 34x - 56$
12. $16p^2 - 24p + 9$
13. a. $6w - 10$
    b. $2w^2 - 5w$
    c. 38 ft, 88 ft^2
14. $(a + 8)(a + 2)$
15. $5(y - 2)(y + 2)$
16. $(2s - 1)(2s + 1)$
17. $b(3b - 10)(b - 1)$
18. $(2x^2 + 1)(5x - 6)$
19. $(4t - 3)^2$
20. $x = 0, 3$
21. $x = 2, -2, 1$
22. $x = 2, \frac{1}{3}$
23. $m = 0, 2$
24. $r = -\frac{1}{4}, -3$
25. $b = 10, 1$
26. a. x^3 ft^3
    b. $6x^2$ ft^2
    c. $(x + 1)^3$ or $(x^3 + 3x^2 + 3x + 1)$ ft^3
    d. $24x^2$ ft^2
27. Sample answer: $(x + 3)(x + 2)(x - 1)$
28. a. $6x^2 + 20x + 30$
    b. 280 ft

Test B

1. $5x^2 - 4x + 5$; 2; 5; trinomial
2. $-5b^3$; 3; -5; monomial
3. $-z^7 + 20z^4 + \frac{2}{3}z$; 7; -1; trinomial
4. $\frac{4}{5}a^5 - \frac{3}{7}a^2$; 5; $\frac{4}{5}$; binomial
5. $-3b^4 - 19b^2 + 25b$
6. $-12x^3 + 4x^2 + 4x$
7. $-7x^2 + 11x - 3$
8. $-12y^3 - 2y^2 - 11y + 5$
9. $-a^3 + 8a^2 - 5a - 50$
10. $-10c^2 - 28c + 6$
11. $6x^4 + 26x^3 + 8x^2$
12. $5p^2 - 30p + 45$
13. a. $(4w + 20)$ cm
    b. $(w^2 + 10w + 16)$ cm^2
    c. 80 cm
14. $5nm(2m^2 - 3n)$
15. $2(2x - y)(x + y)$
16. $3p(p + 10)(p - 7)$
17. $-5(x - 7)(x + 4)$
18. $5x(x + 5)(x - 5)$
19. $(6a - 7x)(b + 2a)$
20. $x = 0, \frac{2}{3}, 5$
21. $x = 3, -3, 1, -1$
22. $p = -6, -2$
23. $m = \frac{1}{3}, -\frac{1}{3}$
24. $x = 0, 4, -1$
25. $x = 2, -2, 1, -1$
26. 3 sec
27. a. $(-w^3 + 40w^2 + 441w)$ in.^{3}
    b. width = 40 in., length = 9 in., height = 49 in. or width = 21 in., length = 28 in., height = 30 in.

Answers

Alternative Assessment

1. a. $(q + 6)(q - 5)$
   b. $(5x + 3)(x - 4)$
   c. $(-2n + 1)(n - 7)$
   d. $(a + 11)(a - 11)$
   e. $(3k + 4)^2$
   f. $(4x^2 + 3)(x - 3)$
   g. $p(p + 7)(p - 7)$

2. a. $k = 8$, $k = -2$, or $k = 7$
   b. $m = 9$
   c. $y = 0$, $y = -4$, or $y = -5$
   d. $x = -10$, $x = 5$, or $x = -5$
   e. $p = 6$, $p = 2$, or $p = -2$

3. a. $(x + 3)$ ft
   b. $(4x + 28)$ ft
   c. $x = 9$
   d. yes; The perimeter is 64 feet. So, the total cost is $176, which is within the budget.

Chapter 8

8.1–8.3 Quiz

1. axis of symmetry: $x = 2$; vertex: $(2, 4)$; decreasing: $x < 2$; increasing: $x > 2$

2. axis of symmetry: $x = -3$; vertex: $(-3, -1)$; increasing: $x < -3$; decreasing: $x > -3$

3. vertical stretch by a factor of 2 with a reflection in the x-axis

4. translation 1 unit up and a vertical stretch by a factor of 4

5. translation 10 units down and a vertical stretch by a factor of 2

6. vertical stretch by a factor of 6

7. vertical shrink by a factor of $\frac{1}{4}$

8. vertical shrink by a factor of $\frac{1}{3}$ with a reflection in the x-axis and translation 4 units down

9. translation 3 units up; $g(x) = 4x^2 + 4$

10. translation 3 units down; $g(x) = \frac{1}{4}x^2 - 8$

11. domain: all real numbers; range: $y \leq 10\frac{1}{8}$

12. domain: all real numbers; range: $y \leq 8\frac{9}{16}$

13. a. 1.58 sec
    b. the water droplet that fell from 22 feet because the other water droplet fell from 40 feet

14. domain: all real numbers greater than or equal to 0; range: $0 \leq y \leq 43.0625$; maximum height: 43.0625 ft

Test A

1. vertical stretch by a factor of 2 and a translation 4 units down

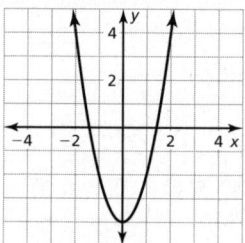

2. reflection in the x-axis and a vertical shrink by a factor of $\frac{1}{3}$

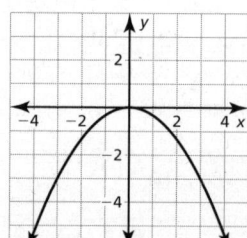

3. vertical stretch by a factor of 2, a translation 3 units left and 5 units down

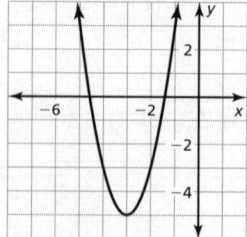

Answers

4. neither **5.** neither **6.** even

7.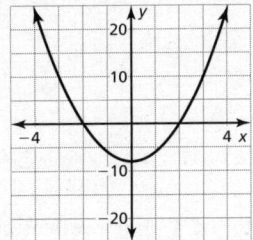

8.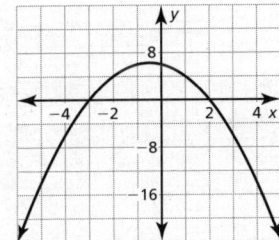

9. $f(x) = 2x^2 + 8x - 10$

10. Sample answer: $f(x) = -x^2 + 2$

11. $f(x) = -\frac{1}{2}x^2 + 2x$

12. linear; $f(x) = -2x + 3$

13. exponential; $f(x) = 2^x$

14. $(5, 0), (-3, 0)$ **15.** $(1, 0), (-7, 0)$

16. $(0, 0), (-4, 0), (4, 0)$

17. $(-5, 0), (-3, 0), (3, 0)$

18. a. domain: all real numbers; range: $y \geq -4$; $(-1, 0), (-5, 0)$
   b. $f(x) = x^2 + 6x + 5$
   c. translation 3 units left and 4 units down
   d. translation 2 units right

19. $(5, 1); x = 5$ **20.** $(3, 0); x = 3$

21. $(6, -137); x = 6$ **22.** $\left(\frac{21}{4}, \frac{385}{8}\right); x = \frac{21}{4}$

23. a. linear; The first differences are equal.
   b. $d(t) = -\frac{1}{2}t + 25$

24. quadratic   **25.** exponential   **26.** linear

Test B

1. reflection in the x-axis, vertical shrink by a factor of $\frac{1}{2}$, and a translation 4 units down

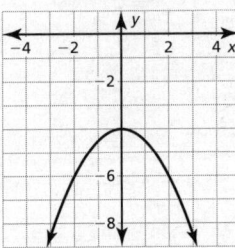

2. vertical stretch by a factor of 2 and a translation 1 unit up

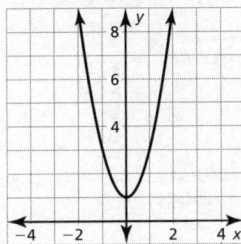

3. reflection in the x-axis, a translation 2 units right and 3 units down

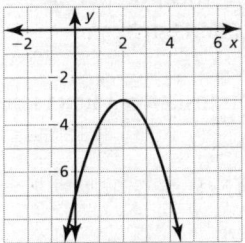

4. neither   **5.** odd   **6.** neither

7.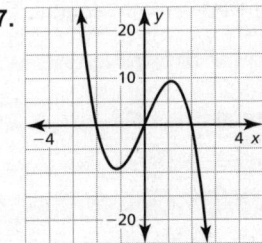

8.

Answers

9. linear; $f(x) = 2x + 12$

10. exponential; $f(x) = \frac{1}{2}^x$

11. quadratic; $f(x) = x^2 - 5$

12. quadratic; $f(x) = 2x^2 + 3$

13. $y = \frac{1}{2}x^2 - \frac{1}{2}x - 3$

14. *Sample answer:* $f(x) = x^2 - 5$

15. $(-6, 0), (1, 0)$ **16.** $\left(\frac{1}{3}, 0\right), \left(-\frac{7}{2}, 0\right)$

17. $(2, 0), (-2, 0), (1, 0), (-1, 0)$

18. $(0, 0), (3, 0), (-3, 0)$

19. a. domain: all real numbers; range: $y \leq 8$; $(5, 0), (1, 0)$
 b. $f(x) = -2x^2 + 12x - 10$
 c. reflection in the *x*-axis, vertical stretch by a factor of 2, translation 3 units right and 8 units up
 d. reflection in the *x*-axis, translation 3 units left and 1 unit down

20. $(-4, -2); x = -4$ **21.** $(1, 0); x = 1$

22. $(6, -8); x = 6$ **23.** $\left(-\frac{27}{2}, -\frac{243}{2}\right); x = -\frac{27}{2}$

24. a. quadratic; The second differences are equal.
 b. $c(t) = 2t^2 + 1$

25. linear   **26.** exponential

Alternative Assessment

1. a.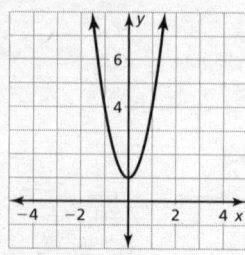

vertical stretch by a factor of 3, followed by a translation 1 unit up

b.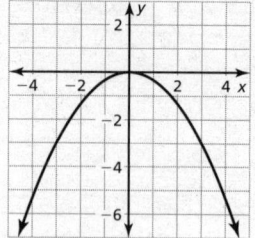

vertical shrink by a factor of $\frac{1}{3}$ and a reflection in the *x*-axis

c.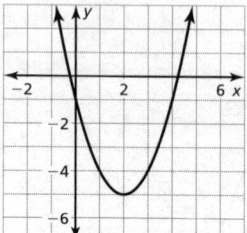

translation 2 units right and 5 units down

d.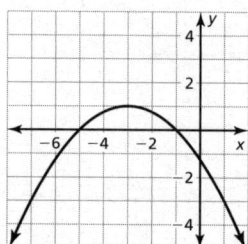

reflection in the *x*-axis and a vertical shrink by a factor of $\frac{1}{4}$, followed by a translation 3 units left and 1 unit up

2. a. domain: all real numbers, range: $y \leq 8$; The zeros are -1 and 3.
 b. vertex: $(1, 8)$; axis of symmetry: $x = 1$
 c. $f(x) = -2x^2 + 4x + 6$
 d. neither
 e. a reflection in the *x*-axis and a vertical stretch by a factor of 2, followed by a translation 1 unit right and 8 units up
 f.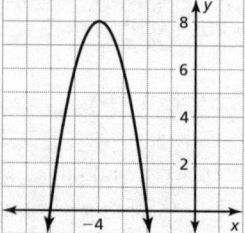

Answers

3. a. $f(x) = -2x^2 - 20x - 42$; Because -7 is a zero, $x + 7$ is a factor. Because -3 is a zero, $x + 3$ is a factor. Setting $f(-4) = 6$ leads to $a = -2$.

b. $f(x) = 2x^2 - 16x$; Because 0 is a zero, x is a factor. Because 8 is a zero, $x - 8$ is a factor. Setting $f(1) = -14$ leads to $a = 2$.

c. *Sample answer:* $f(x) = -x^2 + 2$; Because f is an even quadratic function, $f(x) = ax^2 + d$. Based on the range, the graph opens down, so a is negative, and the maximum is a 2, so $d = 2$.

d. *Sample answer:* $f(x) = x^2 - 6x + 9$; With only two constraints, you can choose the quadratic with $a = 1$. Solve the system of linear equations $9 + 3b + c = 0$ and $36 + 6b + c = 9$ to find $b = -6$ and $c = 9$.

Cumulative Test

1. C 2. D 3. B 4. D
5. C 6. D 7. A 8. B
9. B 10. C 11. A 12. C
13. B 14. D 15. C 16. C
17. B 18. A

19. a. $72x^3y^2$ square units

b. $288x^3y^2$ square units; doubling the length would multiply the area by 2, as would doubling the width

20. a. $V = 32\pi x^3$

b. $V = 30x^3$

21. a. $r = \left(\dfrac{3V}{4\pi}\right)^{1/3}$

b. $r = 6$ cm

22. $\dfrac{46}{15}$

23. $\sqrt[3]{-64}, -2^2 \cdot 2^0$

24. a. 512

b. geometric

c. $a_1 = 1, a_n = 2a_{n-1}$

d. $a_n = 1(2)^{n-1}$

25. a. $y = 175{,}000(1.03)^t$

b. $235{,}185.37

26. $y = 20(1.07)^t$; exponential growth

27. a. 405, 1215, 3645

b. geometric; has a common ratio

c. $a_n = 5(3)^{n-1}$

d. 885,735

28. a. $a_1 = 16{,}384; a_n = \frac{1}{4}a_{n-1}$

b. 64, 16, 4

29. $a_1 = 4, a_n = 2a_{n-1}$; This rule is recursive while the others are explicit.

30. a. $A = 2w^2$

b. $A = 2w^2 + 18w + 36$

c. $A = 18w + 36$

d. 252 square feet

31. a. $A = x^2 + 40x + 400$ square feet

b. 625 square feet; 225 square feet

32. a. 3 sec

b. 147 ft

33. linear:

x	0	1	2	3
y	5	7	9	11

$y = 5x + 3$, graph of line passing through origin

exponential:

x	0	1	2	3
y	3	6	12	24

$y = 5^x$, graph of exponential function

quadratic:

x	0	1	2	3
y	-3	-2	1	6

$y = x^2 + 4$

34. width $= 1$ in., length $= 23$ in.

Answers

Chapter 9

9.1–9.3 Quiz

1. $9x\sqrt{5x}$
2. $-4\sqrt[3]{4}$
3. $\dfrac{\sqrt{6}}{3}$
4. $2\sqrt[3]{\dfrac{2}{7x^2}}$
5. $\dfrac{12 - 4\sqrt{2}}{7}$
6. $-16\sqrt{2}$
7. $x = 3, x = 4$
8. $x = 2$

9.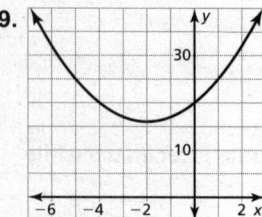
   no real solutions

10.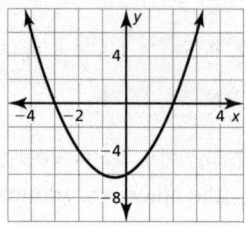
    $x = -3, x = 2$

11. $x = \pm 3\sqrt{3}$
12. $x = \pm 3$
13. $x = -\dfrac{11}{3}$ or $x = \dfrac{7}{3}$
14. two real solutions because $d > 0$
15. a. 0.25 sec and 2.5 sec
    b. 2.86 sec

Test A

1. $5m^3\sqrt{5m}$
2. $\dfrac{6x^2}{11}$
3. $\dfrac{2y^2\sqrt{2y}}{9x^3}$
4. $\dfrac{2\sqrt{6}}{3}$
5. $2 - \sqrt{3}$
6. $-5\sqrt{6} + 4\sqrt{5}$
7. $x = -4, 1$
8. $x = -3, -2$
9. $x \approx 1.2, 5.8$
10. $x \approx -5.4, 0.9$
11. $x = -2, 2$
12. $x = -2, 4$
13. $x = -7, 1$
14. $x \approx -1.7, 5.7$
15. $y = (x + 1)^2$
16. $y = -(x + 4)^2 + 1$
17. $x = -12, 12$
18. $x = 2, 4$
19. $x \approx -1.6, 4.1$
20. $x = -6, \dfrac{2}{3}$
21. $(5, 14), (-1, 8)$
22. $(-0.41, 9), (2.41, 9)$
23. $(-1, 1), (1, 1)$
24. no solution
25. a. 1.875 sec
    b. 1.5625 sec
    c. 3.125 sec
26. a. $8x^2 + 50x + 72$
    b. 130 in.^{2}
    c. $x = 1.5$
27. a. $t \approx 1.07$ sec
    b. 12.5625 ft

Test B

1. $-5xy^2\sqrt[3]{2xy^2}$
2. $\dfrac{a\sqrt{3a}}{3}$
3. $\dfrac{y^3\sqrt{2}}{2x^2}$
4. $-\dfrac{7\sqrt{3}}{6}$
5. $\dfrac{3\sqrt{x} - 3x}{x - x^2}$
6. $-11x\sqrt{5}$
7. $x = -1, 1, 0$
8. $x = -4$
9. -7; no real roots
10. 0; 1 real root
11. -23; no real roots
12. $x = -3, 3$
13. $x = \dfrac{7}{2}, \dfrac{5}{2}$
14. $x \approx 1.54, 8.46$
15. $x = -5, 2$
16. $y = -1(x - 2)^2 + 1$
17. $y = \left(x - \dfrac{5}{2}\right)^2 - \dfrac{65}{4}$
18. $x = -12, 1$
19. $t = -\dfrac{7}{2}, -1$
20. $x = -4, 4$
21. $a \approx -0.38, 1.16$
22. $(5, -10), (2, -10)$
23. $(2, 3), \left(-\dfrac{5}{2}, \dfrac{21}{4}\right)$
24. $(-2, 1), (1, 4)$
25. $(-2, -3), \left(\dfrac{4}{3}, \dfrac{11}{3}\right)$

Answers

26. a.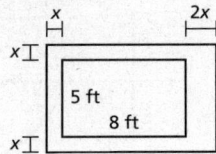

   b. $(3x + 8)(2x + 5)$ or $6x^2 + 31x + 40$

   c. 6 ft

   d. 86 ft

27. a. 0.5 sec

   b. 374 ft

   c. 5.33 sec

28. a. $(w + 2)(2w + 7)$ or $2w^2 + 11w + 14$

   b. $w \approx 10.76$ in.

Alternative Assessment

1. a. $x = -5$ or $x = 2$; *Sample answer:* factoring; The trinomial is easily factorable.

   b. $x = \dfrac{-5 \pm \sqrt{97}}{4}$; *Sample answer:* Quadratic Formula; The trinomial is not factorable and $b \neq 0$.

   c. $x = \pm 12$; *Sample answer:* square roots; $b = 0$

   d. $x = -4 \pm 2\sqrt{7}$; *Sample answer:* Complete the square; there is an even middle term and $a = 1$.

   e. $x = \dfrac{6}{7}$ or $x = -1$; *Sample answer:* factoring; The trinomial is easily factorable.

   f. $x = 1$ or $x = -\dfrac{1}{5}$; *Sample answer:* square roots; The equation is of the form $(\ )^2 = d$.

2. a. $(1, 9)$ and $(-2, -3)$

   b. $\left(\dfrac{5}{4}, -4\right)$ and $(-1, -4)$

   c. $(-3.8285, 2.0003)$

3. a. *Sample answer:* $a = 1$, $b = -5$, and $c = -14$

   b. *Sample answer:* $a = 1$, $b = -6$, and $c = 9$

   c. *Sample answer:* $a = -1$, $b = 3$, and $c = -8$

   d. *Sample answer:* $a = 1$, $b = 1$, and $c = -6$

   e. *Sample answer:* $a = 1$, $b = 0$, and $c = -3$

Chapter 10

10.1–10.2 Quiz

1. $x \geq 6$ **2.** $x \geq 0$ **3.** $x \leq 2$

4. translation 7 units up

5. translation 3 units to the right

6. reflection in the x-axis, a translation 5 units right and 2 units up

7. translation 4 units left

8. reflection in the x-axis, vertical stretch by a factor of 2, and a translation 4 units down

9. reflection in the y-axis and a translation 2 units left

10. $k = 2$ **11.** $a = -1$

12. a.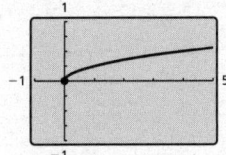

   domain: $h \geq 0$; range: $t \geq 0$

   b. 3003.04 ft

13. 8.03 m/sec

Test A

1. $x \geq -3$ **2.** $x \leq 0$ **3.** $x \geq 2$

4. 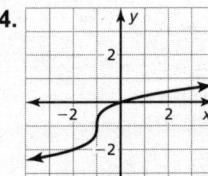   **5.**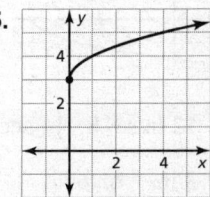

6. 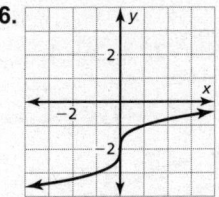   **7.**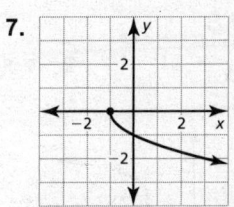

8. reflection in the x-axis, a vertical stretch by a factor of 2, and a translation 3 units right and 4 units up

9. vertical shrink by a factor of $\dfrac{1}{2}$, a reflection in the y-axis, and a translation 3 units right and 1 unit down

Answers

10. $y = 16$ **11.** $x = 16$ **12.** $x = 50$

13. $a = 245.5$ **14.** $x = 64$ **15.** $b = -10$

16. $k = 5, 6$ **17.** $p = 90$ **18.** $y \geq 0$

19. $y \leq 2$

20. $g(x) = -\frac{1}{2}x + 2$ **21.** $g(x) = \frac{1}{3}x - \frac{1}{3}$

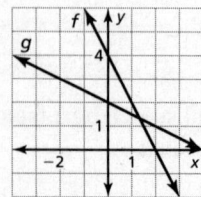 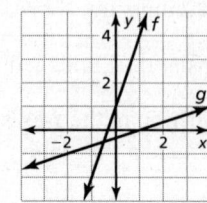

22. $g(x) = \sqrt{x - 4}$; $x \geq 4$

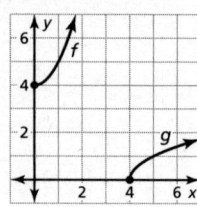

23. $g(x) = -\sqrt{-\frac{1}{2}x + 3}$; $x \leq 6$

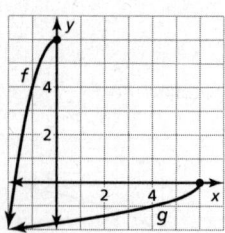

24. a. 12 m/s **b.** 16 sec

25. a. 3 in. **b.** 6 in.

Test B

1. $x \geq -1$ **2.** $x \leq 3$ **3.** $x \geq -4$

4. 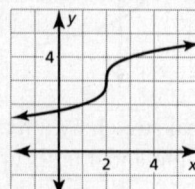   **5.**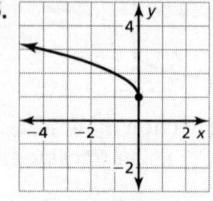

6. 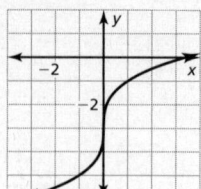   **7.**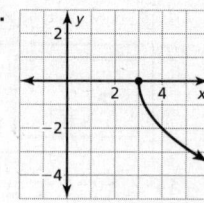

8. vertical shrink by a factor of $\frac{1}{3}$, and a translation 1 unit left and 5 units up

9. reflection in the *x*- and *y*-axes, a vertical stretch by a factor of 2 and a translation 4 units right and 2 units up

10. $n = 1$ **11.** $x = 16$ **12.** $n = 10$

13. $r = 7, 8$ **14.** no solution **15.** $n = 6, 2$

16. $a = -2$ **17.** $m = -9$ **18.** $y \geq 3$

19. $y \leq -5$

20. $g(x) = -x + 5$ **21.** $g(x) = \frac{1}{4}x - \frac{1}{2}$

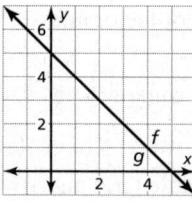 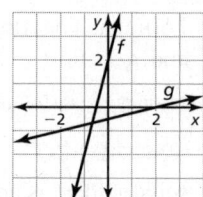

22. $g(x) = -\sqrt{x + 3}$; $x \geq -3$

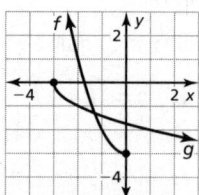

23. $g(x) = \sqrt{2x + 8}$; $x \geq -4$

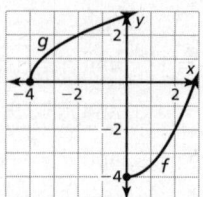

Answers

24. a. 39.72 ft

  b. The time it takes to complete a full swing decreases.

25. a. 4.57 in.

  b. no; volume ≈ 381.70 in.^{3}

Alternative Assessment

1. a.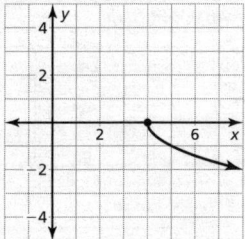

domain: $x \geq 4$; range: $y \leq 0$; reflection in the x-axis and a translation 4 units right

b.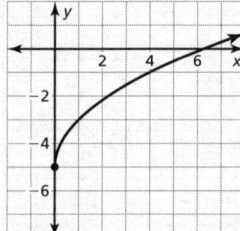

domain: $x \geq 0$; range: $y \geq -5$; vertical stretch by a factor of 2 and a translation 5 units down

c.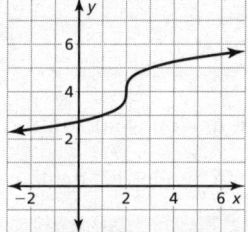

domain: all real numbers; range: all real numbers; translation 2 units right and 4 units up

d.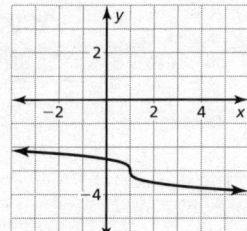

domain: all real numbers; range: all real numbers; reflection in the x-axis and a vertical shrink by a factor of $\frac{1}{2}$, followed by a translation 1 unit right and 3 units down

2. a. $x = 25$

  b. $x = \frac{35}{4}$

  c. $x = -2$

  d. $x = 5$ and $x = 9$

  e. no solution

3. a. $f^{-1}(x) = \dfrac{x + 2}{7}$; domain: all real numbers; range: all real numbers

  b. $f^{-1}(x) = 9x^2 - 36x + 40$; domain: $x \geq 2$; range: $y \geq 4$

  c. $f^{-1}(x) = \sqrt{-2x} - 3$; domain: $x \leq 0$; range: $y \geq -3$

  d. $f^{-1}(x) = (x + 4)^3$; domain: all real numbers; range: all real numbers

  e. $f^{-1}(x) = -2\sqrt{x + 1}$; domain: $x \geq -1$; range: $y \leq 0$

Chapter 11

11.1–11.3 Quiz

1. mean = $6\frac{1}{8}$, median = $6\frac{1}{8}$, no mode; the mean, because the data distribution is symmetric.

2. mean = $1522.\overline{2}$, median = 1400, mode = 1400; the median, because the data distribution is skewed.

3. Female: range = 6, standard deviation = 2.24; Male: range = 7, standard deviation = 2.58; The measures of spread for female and male are similar.

4. Juniors: range = 16, standard deviation = 4.92; Seniors: range = 20, standard deviation = 6.03; The measures of spread for juniors are smaller than the measures of spread for seniors.

5. a. mean = 258, median = 225, no mode, range = 329, standard deviation = 98.59

  b. 480; the mean and median will both decrease after removing the outlier, but there is still no mode.

  c.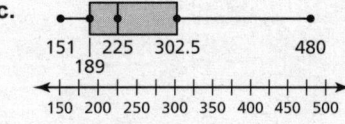

  The data is skewed to the right.

Answers

6. a.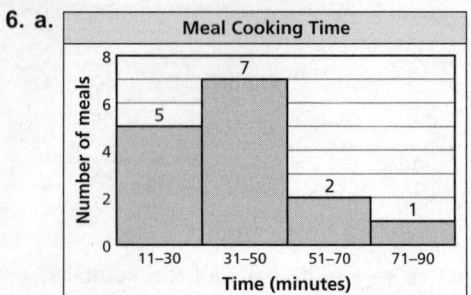

b. median

Test A

1. mean: 4, median: 3.5, mode: 3; median
2. mean: 11.5, median: 11, mode: 11; median
3. 2
4. 5
5. 6
6. 9
7. 11
8. 8
9.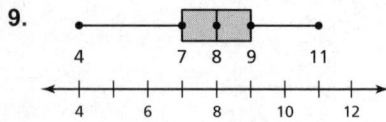
10.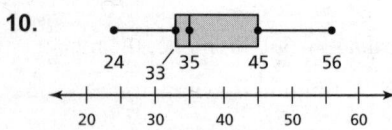
11. **a.** $17; The outlier pulls the mean to the right, making it greater. The median also increases, but the mode is not affected.
    b. *Sample answer:* That student could have more experience or training at the job he or she is working than the others in the survey.
12. science: range = 34, standard deviation = 11.2; math: range = 23, standard deviation = 9.3; The measures of spread for science are larger than the measures of spread for math.
13. freshmen: range = 16, standard deviation = 5.1; seniors: range = 12, standard deviation = 4.1; The measures of spread for freshmen are larger than the measures of spread for seniors.
14. quantitative
15. qualitative
16. quantitative
17. qualitative
18. $x = 10$
19. $x = 3$
20. **a.** Company A: left skew; Company B: right skew
    b. Company A; because company A has a smaller range than company B.
    c. Company B
21. **a.**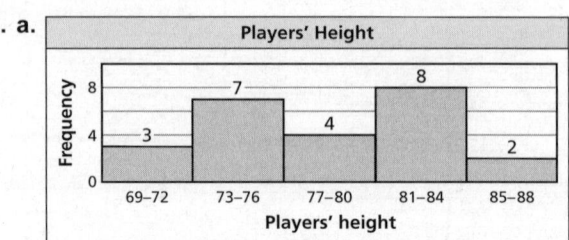

    b. mean; The data is relatively symmetric.
22. *Sample answer:* line graph; The data changes with time.
23. *Sample answer:* circle graph; The data is categorical and out of 100.
24. **a.**

	Male	Female	Total
Pizza	95	109	204
Chicken	94	52	146
Total	189	161	350

   b. 67.7%

Test B

1. mean: 5.6, median: 5.5, mode: 5 and 7; mean
2. mean: 14.5, median: 13.5, mode: 12; median
3. 4
4. 8
5. 14
6. 14
7. 18
8. 16
9.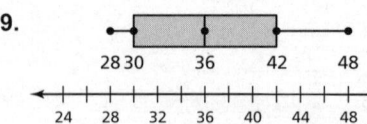
10.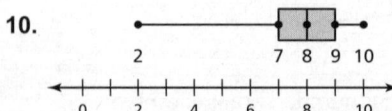

Answers

11. a. 45; The outlier pulls the mean to the left, making it smaller. The median also decreases slightly, but the mode is not affected.

   b. *Sample answer:* That student did not study for the test.

12. Period 2: range = 8, standard deviation = 2.6; Period 4: range = 17, standard deviation = 5.1; The measures of spread for Period 2 are smaller than the measures of spread for Period 4.

13. male: range = 8, standard deviation = 2.6; female: range = 9, standard deviation = 2.9; The measures of spread for female and male are similar.

14. qualitative   **15.** qualitative

16. quantitative   **17.** quantitative

18. $x = 3$ **19.** $x = 19$

20. a. Student 1: slight skew right; Student 2: skew right

   b. Student 2; smaller range

   c. Student 1

21. a.

Wrestler's Weight histogram: 127–142: 8; 143–158: 7; 159–174: 2; 175–190: 2; 191–206: 1

   b. median; The data is skewed right.

22. a.

	Male	Female	Total
Yes	100	89	189
No	71	90	161
Total	171	179	350

   b. 47.1%

Alternative Assessment

1. a. The data distribution is skewed, so use the median and the five-number summary.

   b. The data distribution is symmetric, so use the mean and the standard deviation.

   c. The data distribution is symmetric, so use the mean and the standard deviation.

   d. The data distribution is skewed, so use the median and the five-number summary.

   e. The data distribution is symmetric, so use the mean and the standard deviation.

   f. The data distribution is skewed, so use the median and the five-number summary.

2. a. The mean is 9.58, the median is 9.6, the mode is 9.6, the range is 3.5, and the standard deviation is 0.89.

   b. The mean is 9.98, the median is 10, the mode is 10, the range is 3.5, and the standard deviation is 0.89.

   c. *Sample answer:* box-and-whisker plot; There is a relatively large gap between the outlier 7.5 and the next data value.

3. a.

	Treadmill	Outside	Total
Female	64	27	91
Male	15	52	67
Total	79	79	158

91 females responded. 67 males responded. 158 students were surveyed. 79 students prefer to run on a treadmill. 79 students prefer to run outside.

   b. 22.4%

Cumulative Test

1. C   **2.** A   **3.** A

4. D   **5.** B   **6.** C

7. $(-2, 7), (0, 0), (3, -2), (1, 8), (5, 4)$

8. a. $h = \dfrac{5\sqrt{3}}{2}$ cm   **b.** $P = 12\sqrt{3}$ in.

Answers

9. a.
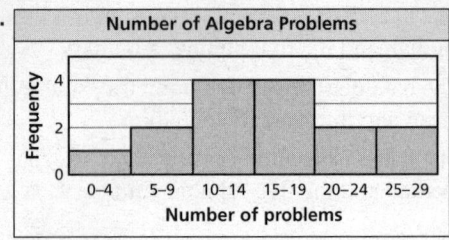

b. symmetric

c. *Sample answer:* The teacher assigns 19 or fewer problems most nights.

10. a. about 225.2 km/h

b. $d = \left(\dfrac{S}{356}\right)^2$

c. about 2.8 km

11. $h = \sqrt{202} - 4 \approx 10.2$ cm

12. a. $y = (x + 7)^2 - 100$

b. minimum at $(-7, -100)$

c. $x = 3$ or -17

13. a. 144 ft

b. 2 sec

c. 5 sec

14. $(2, -1)$ and $(-3, 4)$

15. a. $A \geq 0$

b. 0

c. *Sample answer:* yes; A square with zero area would have a side length of 0.

d. The side length continues to increase as the area increases.

16. a. $r = \sqrt[3]{\dfrac{3V}{4\pi}}$

b. about 9.5 in.

c. no; The radius would be about 4.6 inches.

17. a. $(5, 0)$ and $(-5, 0)$; It takes 5 seconds to descend the hill. The negative intercept would not apply in this situation.

b. 3 sec

18. $2x^2 = -144$; It has no real solutions.

19. a. $s = \pm\sqrt{A}$

b. The side length cannot be negative.

c. $s = 5x\sqrt{2}$ in.

20. a. *Sample answer:* circle graph

b. *Sample answer:*

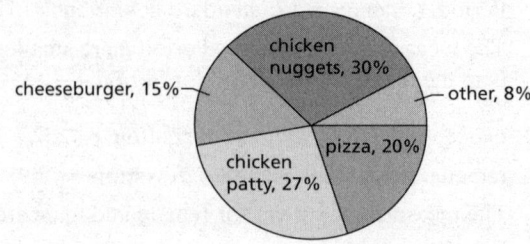

21. square root: $y = x^{1/2} - 4$; $y = -\sqrt{x - 3} + 8$; graph of square root
cube root: $y = x^{1/3} + 7$; $y = \sqrt[3]{x + 4} - 3$; graph of cube root

22. a. The upper quartile should be 11, not 12.

b.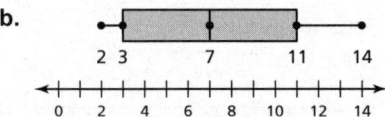

23. a.

	Favorite Past Time	Not Favorite Past Time	Total
Reading	17	24	41
Watching TV	21	19	40
Total	38	43	81

b. about 21%

c. about 26%

Post Course Test Answers

1. $x = -19$ **2.** $x = \dfrac{3}{2}$ **3.** $n = \dfrac{1}{2}, -3$

4. $b = -6$

5. all real numbers except 8

6. $c > 0$ **7.** $\dfrac{n}{3} < 5$ **8.** $y + 10 \geq 17$

9. $x \geq 0$

Answers

10. $x \leq -3$ or $x \geq 0$

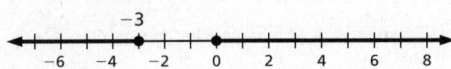

11. $x < -1$ or $x \geq 5$

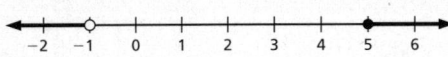

12. $0 < x < 4$

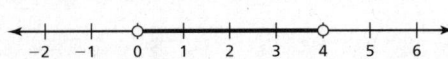

13. yes; linear **14.** yes; nonlinear

15. $y = 2x + 7$ **16.** $y = \frac{4}{3}x$

17. $y + 3 = \frac{8}{3}(x + 1)$ or $y - 5 = \frac{8}{3}(x - 2)$

18. $y - 4 = -\frac{1}{6}(x - 4)$

19.

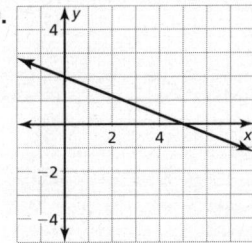

x-int: $(5, 0)$; y-int: $(0, 2)$; slope: $-\frac{2}{5}$

20.

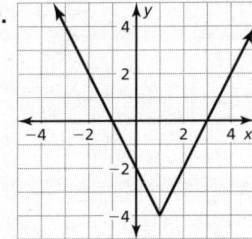

x-int: $(-1, 0), (3, 0)$; y-int: $(0, -2)$

21.

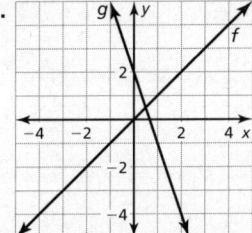

reflection in the x-axis, a vertical stretch by a factor of 3 and a translation 2 units up

22.

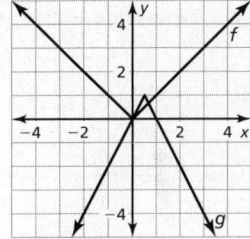

reflection in x-axis, a horizontal shrink by a factor of $\frac{1}{2}$, a translation $\frac{1}{2}$ unit right and 1 unit up.

23.

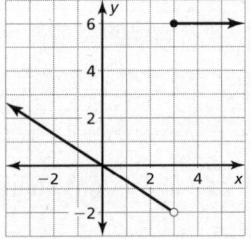

domain: all real numbers; range: $y > -2$

24.

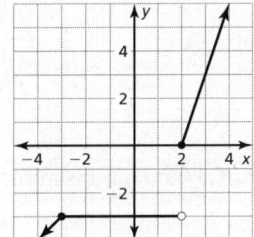

domain: all real numbers; range: $y \leq -3$ or $y \geq 0$

25. $(-2, 3)$ **26.** $(-7, 2)$

27. 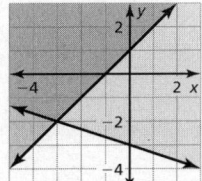 **28.**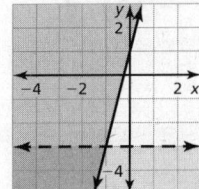

29. 2.15 **30.** 2.45 **31.** 81

32. $\frac{1}{x^4}$ **33.** $\frac{w}{2z^4}$ **34.** $\frac{27}{64a^3b^3}$

35. $r = -\frac{1}{2}$ **36.** $x = \frac{3}{2}$

37. $7x^2 - 2x$; 2; binomial

38. $7x^3 + x - 5$; 3; trinomial

39. $c^2 - 8c + 15$

40. $14a^2 + 41a - 28$ **41.** $4x^2 - 1$

Answers

42. $(b^2 + 1)(b - 3)$ **43.** $-(n + 4)(n - 5)$

44. $(2x - 3)(x - 7)$ **45.** $x = 0, -\frac{1}{3}, 8, -2$

46. $k = -3, -\frac{1}{4}$ **47.** $x = \pm 3\sqrt{2}$

48. $x = 7, 1$ **49.** $x = \dfrac{-1 \pm \sqrt{13}}{3}$

50.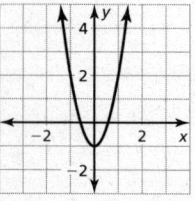

vertical stretch by a factor of 3 and a translation 1 unit down

51.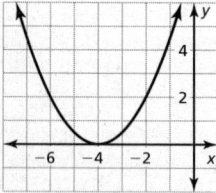

vertical shrink by a factor of $\frac{1}{2}$ and a translation 4 units left

52. $g(x) = -4x - 8$

53. $g(x) = (x + 4)^2 - 3,\ x \geq -4$

54.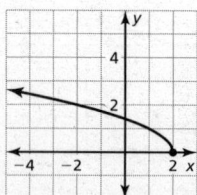

domain: $x \leq 2$; range: $y \geq 0$; reflection in the y-axis and translated 2 units right

55.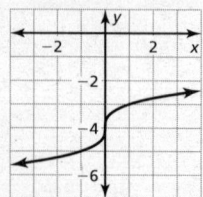

domain: all real numbers; range: all real numbers; translated 4 units down

56. no solution   **57.** $v = 1$ **58.** $x = 4$

59. mean: 9.5; median: 9.75; mode: 10; range: 7

60. mean: 23.7; median: 21.2; mode: none; range: 14.6